Lecture Notes in Mathematics

Edited by A. Dold and B. Eckmann

618

I. I. Hirschman, Jr.
Daniel E. Hughes

Extreme Eigen Values of Toeplitz Operators

Springer-Verlag
Berlin Heidelberg New York 1977

Authors
I. I. Hirschman, Jr.
Washington University
St. Louis, MO 63130/USA

Daniel E. Hughes
Gonzaga University
Spokane, WA 99202/USA

AMS Subject Classifications (1970): 47-02, 47A10, 47A55, 47B35

ISBN 3-540-07147-4 Springer-Verlag Berlin Heidelberg New York
ISBN 0-387-07147-4 Springer-Verlag New York Heidelberg Berlin

Printed in Germany

Printing and binding: Beltz Offsetdruck, Hemsbach/Bergstr.

2141/3140-543210

PREFACE

The asymptotic distribution of the eigen values of finite section Toeplitz operators as the section parameter increases to ∞ has been known ever since the fundamental paper of Szegö, "Ein Grenzwertsatz über die Toeplitzschen Determinanten einer reellen positiven Funktion", Math. Ann. 76, 490-503 (1915). In the last fifteen years interest has been focused on the asymptotic behaviour as the section parameter increases to ∞ of the very large and the very small eigen values. The object of the present exposition is to give a systematic account of one major portion of this subject, incorporating recent advances and discoveries.

TABLE OF CONTENTS

		Page
CHAPTER I. Introduction		
1.	The first eigen value problem	1
2.	The second eigen value problem	10
CHAPTER II. Hilbert Space Background-Small Eigen Values		
1.	A perturbation problem	16
2.	The resolvent equation	20
3.	Spectral resolutions	24
4.	Convergence in dimension	28
CHAPTER III. The Fourier Transform Theorem		
1.	Spaces and operators	31
2.	The application of the perturbation theory	39
3.	Convergence of $\underline{S}^{\wedge}(t)^{\frac{1}{2}}$ to $(\underline{S}^{\wedge})^{\frac{1}{2}}$	44
4.	Convergence of $\underline{F}^{\wedge}(t)$ to $\underline{F}^{\wedge}$	49
5.	Convergence of $\underline{S}^{\wedge}(t)^{\frac{1}{2}}\underline{F}^{\wedge}(t)$ to $(\underline{S}^{\wedge})^{\frac{1}{2}}$ on $\underline{M}^{\wedge}$, Part I	54
6.	Convergence of $\underline{S}^{\wedge}(t)^{\frac{1}{2}}\underline{F}^{\wedge}(t)$ to $(\underline{S}^{\wedge})^{\frac{1}{2}}$ on $\underline{M}^{\wedge}$, Part II	57
7.	The asymptotic formula, I	65
8.	The asymptotic formula, II	70
CHAPTER IV. The Fourier Series Theorem		
1.	Spaces and operators	74
2.	Application of the perturbation theory	78
3.	Convergence of $\underline{S}^{\wedge}(t)^{\frac{1}{2}}$ to $(\underline{S}^{\wedge})^{\frac{1}{2}}$	83
4.	Convergence of $\underline{F}^{\wedge}(t)$ to $\underline{F}^{\wedge}$	83
5.	Convergence of $\underline{S}^{\wedge}(t)^{\frac{1}{2}}\underline{F}^{\wedge}(t)$ to $(\underline{S}^{\wedge})^{\frac{1}{2}}\underline{F}^{\wedge}$, Part I	92
6.	Convergence of $\underline{S}^{\wedge}(t)^{\frac{1}{2}}\underline{F}^{\wedge}(t)$ to $(\underline{S}^{\wedge})^{\frac{1}{2}}$ on $\underline{M}^{\wedge}$, Part II	93
7.	The asympotic formula, I	93
8.	The asympotic formula, II	96

TABLE OF CONTENTS

	Page
CHAPTER V. Hilbert Space Theory - Large Eigen Values	
1. A perturbation theorem	97
2. Convergence in dimension	102
CHAPTER VI. The Fourier Series and Fourier Transform Theorems	
1. Spaces and operators	105
2. Further operators	110
3. $\underline{F}^{\wedge}(t)$ and $\underline{F}^{\wedge}(t)^*$	113
4. $\underline{U}^{\wedge}(t)\underline{F}^{\wedge}(t)$ and $\underline{U}^{\wedge}\underline{F}^{\wedge}$, $\underline{V}^{\wedge}(t)\underline{F}^{\wedge}(t)$ and $\underline{V}^{\wedge}\underline{F}^{\wedge}$	118
5. $\underline{U}^{\wedge}(t)$ and $\underline{U}$, $\underline{V}^{\wedge}(t)$ and $\underline{V}^{\wedge}$.	121
6. $\underline{N}^{\wedge}(t)$	122
7. The asympotic formula	125
8. The Fourier transform	134
9. Asymptotic formulas for the large eigen value problem continued	137
BIBLIOGRAPHY	139
INDEX OF SYMBOLS	142

Chapter I

INTRODUCTION

§1. The First Eigen Value Problem

This paper is devoted to the study of two problems concerning extreme eigen values of Toeplitz operators. It is partly a very systematic exposition (self-contained apart from a knowledge of basic harmonic analysis and Hilbert space theory) of results previously known in a slightly less general form and partly a presentation of new results.

In order to explain the first of these problems we recall Szegö's fundamental theorem on the global distribution of the eigen values of Toeplitz operators. Let $\underline{T}$ be the real numbers $\underline{R}$ modulo 2π and let $L^2(\underline{T})$ be the Hilbert space of those complex Lebesgue measurable functions $u^\wedge(\theta)$ on $\underline{T}$ for which $\|u^\wedge\| < \infty$ where

$$\|u^\wedge(\theta)\| = \langle u^\wedge | u^\wedge \rangle^{1/2},$$

$$\langle u^\wedge | v^\wedge \rangle = \frac{1}{2\pi}\int_{\underline{T}} u^\wedge(\theta)\overline{v^\wedge(\theta)}\, d\theta.$$

Given a real function $f^\wedge(\theta)$ in $L^1(\underline{T})$ let M be the operator

$$Mu^\wedge\cdot(\theta) = f^\wedge(\theta)u^\wedge(\theta)$$

where the domain of M consists of these functions $u^\wedge(\theta) \in L^2(\underline{T})$ for which $|f^\wedge(\theta)u^\wedge(\theta)|^2 \in L^1(\underline{T})$. It is easy to see that M is self-adjoint and that its spectrum is equal to the essential range of $f(\theta)$.

Let $\underline{Z}$ be the integers and let $L^2(\underline{Z})$ be the Hilbert space of those complex functions $u(k)$ on $\underline{Z}$ for which $\|u\| < \infty$ where

$$\|u\| = \langle u|\,u\rangle^{1/2} < \infty,$$

$$\langle u|\,v\rangle = \sum_{k \in \underline{Z}} u(k)\overline{v(k)}.$$

We denote by φ the Fourier transform

$$\varphi : u \to u^\wedge(\theta) = \sum_{\underline{Z}} u(k)e^{ik\theta} \qquad \theta \in \underline{T}$$

which maps $L^2(\underline{Z})$ unitarily onto $L^2(\underline{T})$, and we denote by φ^{-1} the inverse Fourier transform

$$\varphi^{-1} : u^\wedge \to u\cdot(k) = \frac{1}{2\pi}\int_{\underline{T}} u^\wedge(\theta)e^{-ik\theta}\, d\theta \qquad k \in \underline{Z}.$$

It is not difficult to see that if

$$T = \varphi^{-1}M\varphi$$

then (formally)

$$Tu\cdot(k) = \sum_{j \in \underline{Z}} f(k-j)u(j) \qquad u \in L^2(\underline{Z}),$$

where

$$f(k) = \frac{1}{2\pi}\int_{-\pi}^{\pi} f^\wedge(\theta)e^{-ik\theta}\, d\theta.$$

In a variety of applications it is necessary to consider the truncations of T

$$T^{(n)}u\cdot(k) = \sum_{j=0}^{n} f(k-j)u(j) \qquad k = 0,1,\cdots,n\ ,$$

and to study in detail the behaviour of the spectrum of $T^{(n)}$ as $n \to \infty$. The first and one of the most important results in the area is due to G. Szegö. Suppose that $\{\lambda_k^{(n)}\}_{k=0}^{n}$ are the (necessarily real) eigen values of $T^{(n)}$, and that for $-\infty < a < b < \infty$

$$N[(a,b);\ T^{(n)}] = \{\lambda_k^{(n)} \in (a,b)\}^{\#},$$

where $\{\ \}^{\#}$ is the number of elements in the set $\{\ \}$. If

$$|\{\theta:\ f^{\wedge}(\theta) = a\}|_{\underline{T}} = |\{\theta:\ f^{\wedge}(\theta) = b\}|_{\underline{T}} = 0,$$

where $|\{\ \}|_{\underline{T}}$ is the Lebesgue measure of the set $\{\ \}$ in $\underline{T}$, then Szegö showed that

$$\text{(1)} \qquad \lim_{n\to\infty} N[(a,b);\ T^{(n)}]/(n+1) = \frac{1}{2\pi}\,|\{\theta:\ f^{\wedge}(\theta) \in (a,b)\}|_{\underline{T}}\ .$$

We will have occasion to make use of (1). This striking formula has been generalized in many directions.

We now describe in a special case the first of the problems whose solution is the object of this paper. If $T^{(n)}u = \lambda\, u$ and $\|u\| = 1$, then

$$\lambda = (2\pi)^{-1}\int_{\underline{T}} f^{\wedge}(\theta)|u^{\wedge}(\theta)|^2 d\theta$$

where $u^{\wedge}(\theta) = \sum_1^n u(k)e^{ik\theta}$. Let $\{\lambda_k^{(n)}\}_{k=1}^{n}$ be the eigen values of $T^{(n)}$ written in non-decreasing order. It follows that if $M_1 = \text{ess inf}\ f^{\wedge}(\theta)$,

$M_2 = \text{ess sup } f^\wedge(\theta)$, ($M_1$ may be $-\infty$, and M_2 may be $+\infty$), and if $M_1 < M_2$ then

$$M_1 < \lambda_{n,1} \le \lambda_{n,2} \le \cdots \le \lambda_{n,n} < M_2.$$

It is not difficult to see that

$$\lim_{n\to\infty} \lambda_{n,k} = M_1 \qquad k = 1,2,\ldots,$$

(2)

$$\lim_{n\to\infty} \lambda_{n,n+1-k} = M_2 \qquad k = 1,2,\ldots .$$

We are interested in "how" this convergence takes place -- under suitable assumptions on $f^\wedge(\theta)$. In this section we suppose that $M_1 = 0$, $M_2 < +\infty$, and that there exists $\theta_1 \in T$ such that:

$$f^\wedge(\theta) \sim \sigma|\theta-\theta_1|^{\omega} \quad \text{as} \quad \theta \to \theta_1,$$

where $\sigma > 0$, $\omega > 0$, and where if U is any open set in $\underline{T}$ containing θ_1 then

$$\underset{T\setminus U}{\text{g.l.b.}} |f^\wedge(\theta)| > 0.$$

If in (1) we set $a = \lambda_{n,k}$ and $b = \infty$ then for n large and k fixed we find that "essentially"

$$\frac{k}{n} \sim \frac{1}{2\pi} |\{\theta\colon \sigma|\theta-\theta_1|^{\omega} < \lambda_{n,k}\}|_{\underline{T}},$$

which implies that

(3) $$\lambda_{n,k} \sim \sigma(\pi k)^{\omega}\, n^{-\omega}$$

for k fixed as $n \to \infty$. Unrigorous as this argument is the result it suggests is almost correct since, as we shall show, there exists a sequence of constants

(4) $$0 < \mu_1 \leq \mu_2 \leq \cdots \qquad \lim_{k \to \sigma} \mu_k = \infty$$

depending upon ω such that

(5) $$\lambda_{n,k} \sim \sigma \mu_k n^{-\omega}$$

for k fixed as $n \to \infty$.

In order to give an indication of how (5) is proved it will be convenient to replace $\underline{T}$ by $\underline{R}_2$ (2-dimensional Euclidean space), a setting which has the advantage that in it our problems take on a more typical and general form.

Let Ω be a locally star-shaped set in $\underline{R}_2$ which is of finite measure.

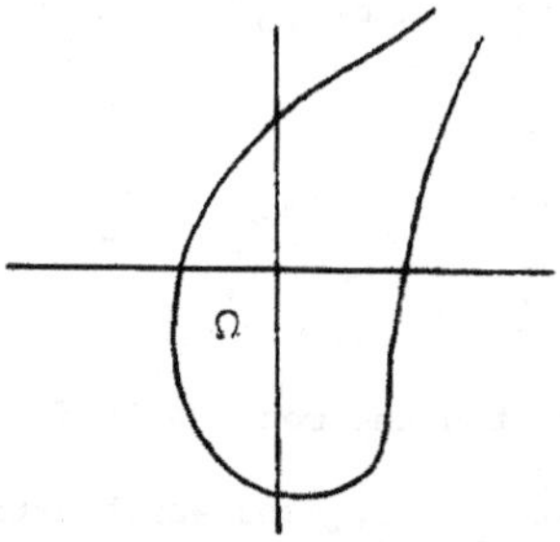

Let $\underset{\sim}{L}$ be the Hilbert space of complex measurable functions on $\underline{R}_2$ determined by the inner product

$$\langle u | v \rangle_{\underset{\sim}{L}} = \int_{\underline{R}_2} u(\underline{\eta}) \overline{v(\underline{\eta})} d\underline{\eta}$$

where $\underline{\eta} = (\eta_1, \eta_2)$. For each $t > 0$ we define a projection $E(t)$ on $\underset{\sim}{L}$ by the formula

$$E(t)u \cdot (\underline{\eta}) = \chi_{t\Omega}(\underline{\eta}) u(\underline{\eta}).$$

Here $\chi_{t\Omega}$ is the characteristic function of the set $t\Omega$. Let $\underset{\sim}{L}^\wedge$ be the Hilbert space of complex measurable functions on $\underline{R}_2$ determined by the inner product

$$\langle u^\wedge | v^\wedge \rangle_{L^\wedge} = (2\pi)^{-2} \int_{\underline{R}_2} u^\wedge(\underline{\xi})\overline{v^\wedge(\underline{\xi})}\,d\underline{\xi}$$

where $\underline{\xi} = (\xi_1,\xi_2)$. Let φ be the unitary mapping $\varphi : u \to u^\wedge$ of $\underset{\sim}{L}$ onto $\underset{\sim}{L}^\wedge$ formally defined by

$$u^\wedge(\underline{\xi}) = \int_{\underline{R}_2} u(\underline{\eta})e^{i\underline{\xi}\cdot\underline{\eta}}\,d\underline{\eta}$$

where $\underline{\xi}\cdot\underline{\eta} = \xi_1\eta_1 + \xi_2\eta_2$. If we set

$$E^\wedge(t) = \varphi E(t)\varphi^{-1}$$

then

$$E^\wedge(t)\cdot u^\wedge\cdot(\underline{\xi}) = \int_{t\Omega} u(\underline{\eta})e^{i\underline{\xi}\cdot\underline{\eta}}\,d\underline{\eta} \qquad u = \varphi^{-1}u^\wedge.$$

Let $f^\wedge(\underline{\xi})$ be a positive bounded measurable function on $\underline{R}_2$ which takes on the value 0 only at $\underline{\xi} = \underline{0}$, and which satisfies

$$\text{g.l.b.}\{f^\wedge(\underline{\xi}) : |\underline{\xi}| > \delta\} > 0$$

for every $\delta > 0$. We further assume that we are given a constant $\omega > 0$, and a positive measurable function $\Phi(\underline{\xi})$ homogeneous of degree 0, bounded away from 0 and ∞, such that

(6) $$\lim_{\underline{\xi}\to\underline{0}} |f^\wedge(\underline{\xi})|\underline{\xi}|^{-\omega} - \Phi(\underline{\xi})| = 0.$$

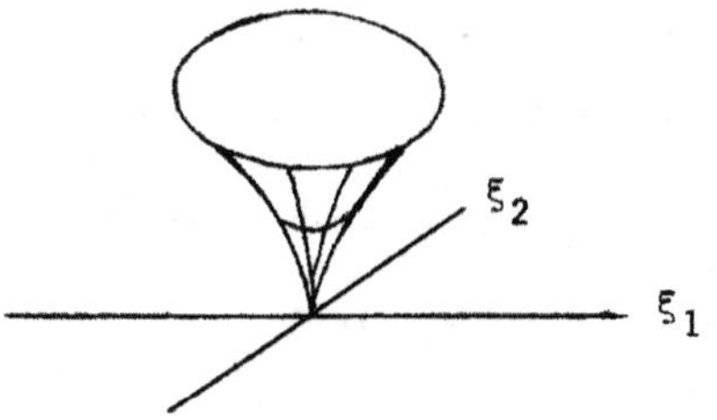

$$\hat{f}(\underline{\xi}) \quad \text{near} \quad \underline{\xi} = \underline{0}.$$

The operator

$$\hat{T}\hat{u}(\underline{\xi}) = \hat{f}(\underline{\xi})\hat{u}(\underline{\xi})$$

is bounded, self-adjoint and positive on $\underset{\sim}{\hat{L}}$.

From $\hat{E}(t)$ and $\hat{T}$ we form the Toeplitz operator

$$\hat{T}_E(t) = \hat{E}(t)\hat{T}\hat{E}(t)\big|_{\hat{E}(t)\underset{\sim}{\hat{L}}},$$

the restriction of $\hat{E}(t)\hat{T}\hat{E}(t)$ to the subspace $\hat{E}(t)\underset{\sim}{\hat{L}}$. We will show that as $t \to +\infty$ the bottom of the spectrum of $\hat{T}_E(t)$ becomes discrete so that it is possible to speak of the smallest eigen value $\lambda(1,t)$ of $\hat{T}_E(t)$, the next smallest $\lambda(2,t)$, etc., provided t is sufficiently large. Note that, as is usual, eigen values are repeated according to their multiplicities. We will further show that there exists a sequence

$$0 < \mu(1) \leq \mu(2) \leq \cdots, \qquad \mu(k) \to \infty \quad \text{as} \quad k \to \infty,$$

depending only upon Ω, ω, and $\Phi(\underline{\xi})$, such that for $k = 1, 2, \cdots$

$$\lambda(k,t) \sim t^{-\omega}\mu(k) \qquad \text{as} \quad t \to \infty. \tag{7}$$

This is proved by renormalizing the operators $\hat{T}_E(t)$ in such a way that the new operators converge as $t \to \infty$.

Let $\chi(t)$ be the unitary transformation on $\underset{\sim}{L}^{\wedge}$ defined by

$$\chi(t)u^{\wedge}\cdot(\underline{\xi}) = t^{-1}u^{\wedge}(t^{-1}\underline{\xi})$$

where $t\underline{\xi} = (t\xi_1, t\xi_2)$, and let $\chi(t)^{-1}$,

$$\chi(t)^{-1}u^{\wedge}\cdot(\underline{\xi}) = tu^{\wedge}(t\underline{\xi}),$$

be its inverse. We set

$$\chi(t)t^{\omega}T_E^{\wedge}(t)\chi(t)^{-1}\big|_{E^{\wedge}(t)\underset{\sim}{L}^{\wedge}} = S_F^{\wedge}(t).$$

It is easy to verify that

$$\chi(t)E^{\wedge}(t)\chi(t)^{-1} = F^{\wedge}$$

where

$$F^{\wedge}u^{\wedge}\cdot(\underline{\xi}) = \int_{\Omega} u(\underline{\eta})e^{i\underline{\xi}\cdot\underline{\eta}}d\underline{\eta} \qquad u = \varphi^{-1}u^{\wedge},$$

and that

$$\chi(t)t^{\omega}T^{\wedge}\chi(t)^{-1} = S^{\wedge}(t)$$

where

$$S^{\wedge}(t)u^{\wedge}\cdot(\underline{\xi}) = t^{\omega}f^{\wedge}(t^{-1}\underline{\xi})u^{\wedge}(\underline{\xi}).$$

Consequently

$$S_F^{\wedge}(t) = F^{\wedge}S^{\wedge}(t)F^{\wedge}\big|_{F^{\wedge}\underset{\sim}{L}^{\wedge}}.$$

It follows from (6) that as $t \to \infty$ $S^{\wedge}(t)$ converges to $S^{\wedge}$ in a sense which we will not make precise here where

$$S^{\wedge}u^{\wedge}\cdot(\underline{\xi}) = |\underline{\xi}|^{\omega}\phi(\underline{\xi})u^{\wedge}(\underline{\xi}).$$

Formally, $S_F^{\wedge}(t)$ would seem to converge to

$$(8) \qquad S_F^{\wedge} = F^{\wedge}S^{\wedge}F^{\wedge}\big|_{F^{\wedge}\underset{\sim}{L}^{\wedge}} .$$

Assuming that the spectrum of $S_F^{\wedge}$ is

$$0 < \mu(1) \leq \mu(2) \leq \cdots , \qquad \lim_{k \to \infty} \mu(k) = \infty,$$

It would be reasonable to expect that

$$\lim_{t \to \infty} t^{\omega}\lambda(k,t) = \mu(k) \qquad k = 1,2,\cdots .$$

In fact, this is what happens. An initial difficulty is that the operator $S_F^{\wedge}$ as defined in (8) is only symmetric and not self-adjoint. If, however, we replace this operator by its Friedrichs' extension then the program sketched above can be carried through, although the actual details are a good deal more sophisticated (and therefore more interesting) than the presentation above suggests.

The problem we have discussed was first treated in the case $\omega = 2$ by Kac, Murdock, and Szegö in [4]. It was subsequently studied by Widom [11] - [14], and Parter [7] - [9] in an intertwining series of investigations culminating in [14], a paper in which Widom obtained in n-dimensions, both in the Fourier transform and Fourier series cases theorems which contain the results just described for $n = 2$ in the Fourier transform case. Somewhat before [14] appeared Kesten treated the Fourier series case in dimension 1 under the more general assumption that f has finitely many zeros of the same order and finitely many zeros of lower order. In Chapters II - IV we have extended Widom's methods so that they apply to this more general case. The perturbation theory of [14], which was put into a more abstract form in [3], is further refined here. In addition various points in Widom's exposition have been reworked resulting in some increase in generality.

§2. The Second Eigen Value Problem

Let $W(\underline{\eta}) \in L^2(\underline{R}_2) \cap L^\infty(\underline{R}_2)$. We define for each $t > 0$ the bounded normal operator $E(t)$ on $\underset{\sim}{L}$ by the formula

(1) $$E(t)u(\underline{\eta}) = W(t^{-1}\underline{\eta})u(\underline{\eta}).$$

It might seem from our earlier discussion that we should take $W(\underline{\eta})$ to be the characteristic function of a suitable set of finite measure, but it is almost as easy to consider the more general case. The adjoint $E(t)^*$ is given by

$$E(t)^* u\cdot(\underline{\eta}) = \overline{W(t^{-1}\underline{\eta})}u(\underline{\eta}).$$

If we set

$$E^\wedge(t) = \omega E(t)\omega^{-1},$$

then

$$E^\wedge(t)u^\wedge\cdot(\underline{\xi}) = \int_{\underline{R}_2} u(\underline{\eta})W(t^{-1}\underline{\eta})e^{i\underline{\xi}\cdot\underline{\eta}}d\underline{\eta} \qquad u = \omega^{-1}u^\wedge$$

$$E^\wedge(t)^* u^\wedge(\underline{\xi}) = \int_{\underline{R}_2} u(\underline{\eta})W(t^{-1}\underline{\eta})e^{i\underline{\xi}\cdot\underline{\eta}}d\underline{\eta}.$$

Let $f^{\wedge}(\underline{\xi})$ be a real measurable function on $\underline{R}_2$ which is bounded from below and which satisfies:

$$\text{l.u.b.}\,\{f^{\wedge}(\underline{\xi}) : |\underline{\xi}| > \delta\} < \infty$$

for each $\delta > 0$. We further assume that we are given a constant ω, $0 < \omega < 2$, and a positive measurable function $\Phi(\underline{\xi})$ homogeneous of degree 0 and bounded away from ∞ such that (in some sense)

(2) $$\lim_{\underline{\xi} \to \underline{0}} \left| f^{\wedge}(\underline{\xi})|\underline{\xi}|^{\omega} - \Phi(\underline{\xi}) \right| = 0.$$

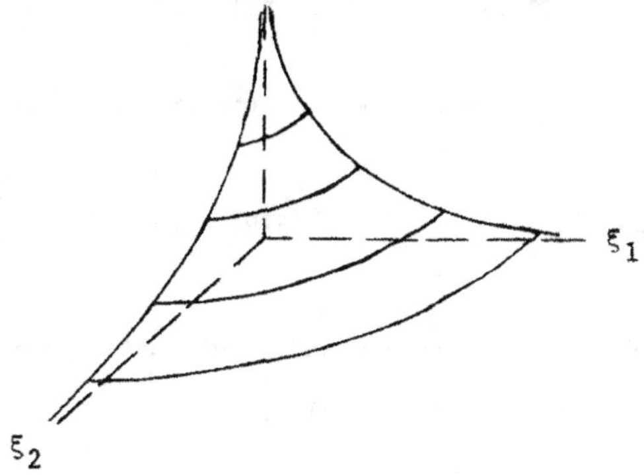

$f^{\wedge}(\underline{\xi})$ near $\underline{\xi} = \underline{0}$

Because $0 < \omega < 2$ and $\Phi(\underline{\xi})$ is bounded it follows that

$$\int_{|\underline{\xi}| \leq 1} |f^{\wedge}(\underline{\xi})|\, d\underline{\xi} < \infty, \quad \text{l.u.b.}_{|\underline{\xi}| > 1} |f^{\wedge}(\underline{\xi})| < \infty.$$

If for $u^{\wedge} \in \underset{\sim}{L}^{\wedge}$ such that $f^{\wedge}(\xi)u^{\wedge}(\xi) \in L^{\wedge}$ we set

$$T^{\wedge}u^{\wedge}(\underline{\xi}) = f^{\wedge}(\underline{\xi})u^{\wedge}(\underline{\xi}),$$

$T^{\wedge}$ is an unbounded self-adjoint operator.

From $E^{\wedge}(t)$ and $T^{\wedge}$ we form the generalized Toeplitz operator

(3) $$T_E^{\wedge}(t) = E^{\wedge}(t)^{*} TE^{\wedge}(t).$$

This definition is, however, unsatisfactory since it does not define $T_E^{\wedge}(t)$ for all $u^{\wedge} \in \underset{\sim}{L}^{\wedge}$, although in fact $T_E^{\wedge}(t)$ is a bounded operator. We therefore proceed as follows. Because $W \in L^2 \cap L^{\infty}$ it follows that

$$\|E^{\wedge}(t)u^{\wedge}\|_{L^{\wedge}} \leq \|W(t^{-1}\underline{\eta})\|_{\infty} \|u^{\wedge}\|_{\underset{\sim}{L}^{\wedge}} ,$$

$$\|E^{\wedge}(t)u^{\wedge}\|_{\infty} \leq \|W(t^{-1}\underline{\eta})\|_{\underset{\sim}{L}} \|u^{\wedge}\|_{\underset{\sim}{L}^{\wedge}} .$$

Using (3) and (4) it is easy to see that the symmetric sesquilinear form

(4) $$(u^{\wedge},v^{\wedge})_t = (2\pi)^{-2} \int_{\underline{R}_2} f^{\wedge}(\underline{\xi})[E^{\wedge}(t)u^{\wedge}\cdot(\underline{\xi})]\overline{[E^{\wedge}(t)v^{\wedge}\cdot(\underline{\xi})]}d\underline{\xi}$$

is bounded for each $t > 0$. By a standard elementary result there exists a bounded self-adjoint operator $T_E^{\wedge}(t)$ such that

$$\langle T_E^{\wedge}(t)u^{\wedge} | v^{\wedge}\rangle_{\underset{\sim}{L}^{\wedge}} = (u^{\wedge},v^{\wedge})_t$$

for all $u^{\wedge},v^{\wedge} \in \underset{\sim}{L}^{\wedge}$. It is this definition of $T_E^{\wedge}(t)$ that we use. It is easy to see that $T_E^{\wedge}(t)u^{\wedge}$ is equal to $E^{\wedge}(t)^{*}TE^{\wedge}(t)u^{\wedge}$ whenever the latter is defined. We will show that as $t \to \infty$ the top of the spectrum of $T_E^{\wedge}(t)$ becomes discrete so that it is possible to speak of the largest eigen value $\lambda^{+}(1,t)$ of $T_E^{\wedge}(t)$, the next largest $\lambda^{+}(2,t)$, etc., provided t is sufficiently large. We will further show that there exists a sequence

$$\mu^{+}(1) \geq \mu^{+}(2) \geq \cdots , \quad \mu^{+}(k) \to 0+ \quad \text{as} \quad k \to \infty,$$

depending only upon W, ω, and Φ, such that for each $k = 1,2,\cdots$

(5) $$\lambda^{+}(k,t) \sim \mu^{+}(k)t^{\omega} \qquad \text{as} \quad t \to +\infty .$$

In outline the demonstration of (5) is almost like that of (7) §1. It is easy to see that

$$\chi(t)E^{\wedge}(t)\chi(t)^{-1} = F^{\wedge}$$

where

$$F^{\wedge}u^{\wedge}(\underline{\xi}) = \int_{\underline{R}_2} W(\underline{\eta})u(\underline{\eta})e^{i\underline{\xi}\cdot\underline{\eta}}d\underline{\eta} \qquad u = \varphi^{-1}u^{\wedge},$$

and that

$$\chi(t)t^{-\omega}T^{\wedge}\chi(t)^{-1} = S^{\wedge}(t)$$

where

$$S^{\wedge}(t)u^{\wedge}(\underline{\xi}) = t^{-\omega}f^{\wedge}(t^{-1}\underline{\xi})u^{\wedge}(\underline{\xi}).$$

It follows from (2) that as $t \to \infty$, $S^{\wedge}(t)$ converges to $S^{\wedge}$ where

$$S^{\wedge}u^{\wedge}\cdot(\underline{\xi}) = |\underline{\xi}|^{-\omega}\,\Phi(\underline{\xi})u^{\wedge}(\underline{\xi}),$$

which suggests that as $t \to \infty$ $S^{\wedge}(t)$ converges to

$$S^{\wedge}_{F} = F^{\wedge}S^{\wedge}F^{\wedge}.$$

Again, $S^{\wedge}_{F}$ is actually defined by means of a sesquilinear form. If the spectrum of $S^{\wedge}_{F}$ is

$$\mu^{+}(1) \geq \mu^{+}(2) \geq \cdots , \mu^{+}(k) \to 0+ \qquad \text{as} \quad k \to \infty,$$

then we should expect that

$$\lim_{t \to \infty} t^{-\omega}\lambda^{+}(k,t) = \mu^{+}(k) \qquad k = 1,2,\cdots$$

and this is indeed the case.

It seems worth stopping to remark that there exists a generalization of the formula of Szegö, (1) of §1, to the present case, due to Kac, Murdock and Szegö, [13], which we give here in a slightly more general formulation. Let $W(\underline{\eta}) \in L^2(\underline{R}_2)$ and let $f^{\wedge}(\underline{\eta}) \in R\ell\, L^1(\underline{R}_2)$.

$$f(\underline{\eta}) = (2\pi)^{-2}\int_{\underline{R}_2} f^{\wedge}(\underline{\eta})e^{i\underline{\xi}\cdot\underline{\eta}}d\underline{\eta}.$$

Then for $T_E(t) = \varphi^{-1}T_E^{\wedge}(t)\varphi$ where $T_E^{\wedge}(t)$ is defined by (3) we have for $a,b \in \underline{R}_2$, $ab > 0$,

$$(6) \qquad \lim_{t \to \infty} N[(a,b);\, T_E(t)]t^{-2} = \frac{1}{(2\pi)^4}\left|\{(\eta,\xi):\, |W(\underline{\eta})|^2 f^{\wedge}(\underline{\xi}) < b\}\right|_{\underline{R}_2 \times \underline{R}_2}$$

whenever

$$\left|\{(\underline{\eta},\underline{\xi}):\, a = K(\underline{\eta})^2 f^{\wedge}(\underline{\xi})\}\right|_{\underline{R}_2 \times \underline{R}_2} = 0,$$

and

$$\left|\{(\underline{\eta},\underline{\xi}):\, b = K(\underline{\eta})^2 f^{\wedge}(\underline{\xi})\right|_{\underline{R}_2 \times \underline{R}_2} = 0.$$

Note that for $u \in L^2(\underline{R}_2)$ we have

$$(7) \qquad T_E(t)u(\underline{\eta}) = \int_{\underline{R}_2} K(t\underline{\eta})f(\underline{\eta}-\underline{\eta}')K(t\underline{\eta}')d\underline{\eta}'.$$

In both the first and second eigen value problems we can replace $\underset{\sim}{L}$ and $\underset{\sim}{L}^{\wedge}$ by $\underset{\sim}{E}$ and $\underset{\sim}{E}^{\wedge}$ where $\underset{\sim}{E}$ is the Hilbert space defined on $\underline{Z}_2$, the two-dimensional lattice group, by the inner product

$$\langle u|v\rangle_{\underset{\sim}{E}} = \sum_{\underline{Z}_2} u(\underline{k})\overline{v(\underline{k})},$$

where $k = (k_1,k_2)$, and $E^\wedge$ is the Hilbert space defined on $\underline{T}_2$, the two dimensional torus group, by the inner product

$$\langle u^\wedge|v^\wedge\rangle_{\underset{\sim}{E}^\wedge} = (2\pi)^{-2}\int_{\underline{T}_2} u^\wedge(\underline{\theta})\overline{v^\wedge(\underline{\theta})}d\underline{\theta}.$$

Chapter II

HILBERT SPACE BACKGROUND - SMALL EIGEN VALUES

1. A Perturbation Problem

Let $\underset{\sim}{H}$ be a Hilbert space with elements f, g, h, etc., inner product $\langle \cdot | \cdot \rangle$ and norm $\| \cdot \|$. Let S be an unbounded self-adjoint operator in $\underset{\sim}{H}$ and for each $t > 0$ let $S(t)$ be a bounded self-adjoint operator in $\underset{\sim}{H}$. Suppose also that F and $F(t)$, $t > 0$, are projections in $\underset{\sim}{H}$. For each t set $S_F(t) = F(t)\, S(t)\, F(t)$. Let its spectral decomposition be

$$S_F(t) = \int \lambda \, d\Psi(t,\lambda) .$$

Formally let $S_F = FSF$ and let its spectral decomposition be

$$S_F = \int \lambda \, d\Psi(\lambda) .$$

The problem we wish to study is that of passing from the convergence of the $S(t)$'s to S and the $F(t)$'s to F to the convergence of the $\Psi(t,\lambda)$'s to $\Psi(\lambda)$. We will show that this can be done in the case $S \geq 0$ and $S(t) \geq 0$, $t > 0$.

The perturbation theory given here, a variant of that developed in [3], is less general but more smoothly applicable to the problem considered in Chapters III and IV.

Throughout this development we assume the Hilbert space $\underset{\sim}{H}$ to be separable. While this is not necessary it makes possible a simpler and more intuitive language. We assume henceforth that:

i. $0 \leq S$ is a self-adjoint operator on $\underset{\sim}{H}$;

ii. F is a projection on $\underset{\sim}{H}$.

We define

$$\underset{\sim}{S} = \{f \in F\underset{\sim}{H}: f \in \underset{\sim}{D}[S^{\frac{1}{2}}]\} .$$

Here $S^{\frac{1}{2}}$ is the unique positive square root of S and $\underset{\sim}{D}[S^{\frac{1}{2}}]$ is its domain. We do not assume that $\underset{\sim}{S}$ is dense in $F\underset{\sim}{H}$ although this is the most interesting special case. Let $\underset{\sim}{M}$ be the closure of $\underset{\sim}{S}$ in $F\underset{\sim}{H}$. $\underset{\sim}{M}$ is a closed subspace of $F\underset{\sim}{H}$ and inherits the structure of a Hilbert space from $\underset{\sim}{H}$. We assert that there exists a self-adjoint transformation S_F on the Hilbert space $\underset{\sim}{M}$ with the properties:

(1) $$\underset{\sim}{D}[S_F] \subset \underset{\sim}{S} ;$$

(2) $$\langle S_F f | g\rangle = \langle S^{\frac{1}{2}} f | S^{\frac{1}{2}} g\rangle$$

for all $f \in \underset{\sim}{D}[S_F]$ and for all $g \in \underset{\sim}{S}$. The construction of S_F with these properties is a special case of the Friedrichs' extension and has long been known, but for the sake of completeness we repeat it here.

In all that follows we will use $\rightarrow$ to indicate strong convergence and $\rightharpoonup$ to indicate weak convergence of elements of $\underset{\sim}{H}$. Similarly $\rightrightarrows$ indicates uniform convergence, $\rightarrow$ strong convergence, and $\rightharpoonup$ weak convergence of operators on $\underset{\sim}{H}$.

Lemma 1a. Let A be a self-adjoint operator on $\underset{\sim}{H}$ and let $h_n \in \underset{\sim}{D}[A]$ $n = 1, 2, \ldots$. If

$$h_n \rightharpoonup h \qquad \text{as } n \rightarrow \infty$$

and if

$$\|Ah_n\| = O(1) \qquad \text{as } n \rightarrow \infty ,$$

then $h \in \underset{\sim}{D}[A]$ and $Ah_n \rightharpoonup Ah$.

This is a special case of Lemma 2a, which is proved later.

For f, $g \in \underset{\sim}{S}$ let us define

$$(f|g) = \langle S^{\frac{1}{2}}Ff | S^{\frac{1}{2}}Fg\rangle + \langle f|g\rangle , \tag{3}$$

$$|||f||| = (f|f)^{\frac{1}{2}} .$$

Lemma 1b. With the definition of inner product and norm given by (3) $\underset{\sim}{S}$ is a Hilbert space.

Proof. It is evident that $\underset{\sim}{S}$ is a pre-Hilbert space. We need only verify that $\underset{\sim}{S}$ is complete. Suppose $f_n \in \underset{\sim}{S}$ $n = 1, 2, \ldots,$ $|||f_n - f_m||| \to 0$ as $n,m \to \infty$. Since $\|f_n - f_m\| \leq |||f_n - f_m|||$ there exists $f \in \underset{\sim}{H}$ such that $\|f - f_n\| \to 0$ as $n \to \infty$; which implies that $\|F(f_n - f)\| \to 0$ as $n \to \infty$. Because $\|S^{\frac{1}{2}}F(f_n - f_m)\| \leq |||f_n - f_m|||$ there exists $g \in \underset{\sim}{H}$ such that $\|S^{\frac{1}{2}}Ff_n - g\| \to 0$ as $n \to \infty$. Applying Lemma 1a with $h_n = Ff_n$ and $A = S^{\frac{1}{2}}$ we see that (since weak and strong limits coincide when both exist) $Ff \in \underset{\sim}{D}[S^{\frac{1}{2}}]$, and that $g = S^{\frac{1}{2}}Ff$. Thus $f \in \underset{\sim}{S}$ and

$$|||f - f_n|||^2 = \|S^{\frac{1}{2}}F(f - f_n)\|^2 + \|f - f_n\|^2 \to 0 \qquad \text{as } n \to \infty .$$

Lemma 1c. There exists a linear transformation W of $\underset{\sim}{M}$ into $\underset{\sim}{S}$ such that $\langle f|g\rangle = (f|Wg)$ for all $f \in \underset{\sim}{S}$, $g \in \underset{\sim}{M}$ and:

i. $\|Wf\| \leq |||Wf||| \leq \|f\|$ for all $f \in \underset{\sim}{M}$;

ii. $\langle Wf|g\rangle = \langle f|Wg\rangle$ for all $f,g \in \underset{\sim}{M}$;

iii. $0 < \langle Wf|f\rangle$ all $f \in \underset{\sim}{M}$.

Proof. For $g \in \underset{\sim}{M}$ fixed $f \to \langle f|g\rangle$ is a linear functional on $\underset{\sim}{S}$ and, since

$$|\langle f|g\rangle| \leq \|f\| \, \|g\| \leq |||f||| \, \|g\| ,$$

there exists a unique element $g' \in \underset{\sim}{S}$, $|||g'||| \leq \|g\|$, such that

$$\langle f|g\rangle = (f|g') \qquad \text{all } f \in \underset{\sim}{S} .$$

Clearly the map $g \to g'$ defines a linear transformation $g' = Wg$ of $\underset{\sim}{M}$ into $\underset{\sim}{S}$ and it is evident that $|||Wg||| \leq \|g\|$ so that i. holds. Suppose that $f, g \in \underset{\sim}{S}$. Then

$$\langle Wf|g\rangle = (Wf|Wg) = \overline{(Wg|Wf)} = \overline{\langle Wg|f\rangle} = \langle f|Wg\rangle$$

so that ii. is valid in this case. By continuity it is valid for $f, g \in \underset{\sim}{M}$. Thus W is a self-adjoint transformation on $\underset{\sim}{M}$. Since

$$\langle Wf|f\rangle = (Wf|Wf) \geq 0 \qquad f \in \underset{\sim}{M}$$

we have $0 \leq W$. To show that $0 < W$ we need only verify that $Wf = 0$ is impossible unless $f = 0$. If $Wf = 0$ then

$$\langle g|f\rangle = (g|Wf) = 0 \qquad \text{all } g \in \underset{\sim}{S} ,$$

but since $\underset{\sim}{S}$ is dense in $\underset{\sim}{M}$ this implies that $f = 0$.

Theorem 1d. There exists a self-adjoint operator S_F on $\underset{\sim}{M}$ satisfying conditions (1) and (2).

Proof. We define

$$S_F = W^{-1} - I .$$

It is evident from this definition that S_F is a self-adjoint operator and that

$$\underset{\sim}{D}[S_F] = \underset{\sim}{D}[W^{-1}] = \underset{\sim}{R}[W] \subset \underset{\sim}{S} ,$$

where $\underset{\sim}{R}[W]$ is the range of W. If $f \in \underset{\sim}{D}[S_F]$ and $g \in \underset{\sim}{S}$ then

$$\langle S_F f | g \rangle = \langle W^{-1} f | g \rangle - \langle f | g \rangle = (f | g) - \langle f | g \rangle ,$$
$$= \langle S^{1/2} f | S^{1/2} f \rangle ,$$

and our proof is complete.

2. The Resolvent Equation

Let A be a closed linear operator on $\underset{\sim}{H}$. It is not assumed that $\underset{\sim}{D}[A]$ is dense in $\underset{\sim}{H}$. A subset $\underset{\sim}{C} \subset \underset{\sim}{D}[A]$ is said to be a core for $\underset{\sim}{A}$ if the set $\{(f,Af): f \in \underset{\sim}{C}\}$ in $\underset{\sim}{H} \times \underset{\sim}{H}$ is dense in the set $\{(f,Af): f \in \underset{\sim}{D}[A]\}$. Let A_t $0 < t < \infty$ and A be closed linear operators in $\underset{\sim}{H}$ and let $C = \{f: A_t f \to Af$ as $t \to \infty\}$. If $\underset{\sim}{C} = \underset{\sim}{D}[A]$ then we say that A is the strong limit of the A_t's ; if $\underset{\sim}{C}$ is a core for A we say that A is the closure of the strong limit of the A_t's .

<u>Lemma 2a</u>. Let A_t $0 < t < \infty$ and A be self-adjoint operators on $\underset{\sim}{H}$ and let A be the closure of the strong limit of A_t as $t \to \infty$. Then if as $t \to \infty$

$$f_t \rightharpoonup f , \ \|A_t f_t\| = O(1) ,$$

we have

$$f \in \underset{\sim}{D}[A] \quad \text{and} \quad A_t f_t \rightharpoonup Af \quad \text{as} \quad t \to \infty .$$

Proof. For the sake of completeness we give the simple proof of this result.

Let $\underset{\sim}{P}$ be the positive real numbers. By a subsequence $\underset{\sim}{P}_1$ of $\underset{\sim}{P}$ we mean a subset $\{t_1, t_2, t_3, \ldots\}$ of $\underset{\sim}{P}$ with $0 < t_1 < t_2 < \ldots$

such that $t_k \to \infty$. By $a_t \to a$ as $t \to \infty$ in $\underset{\sim}{P}_1$ we mean that $\lim_{k\to\infty} a_{t_k} = a$. This notation enables us to dispense with awkward subscripts.

Since by assumption $\|A_t f_t\| = O(1)$, given any subsequence $\underset{\sim}{P}_1$ of $\underset{\sim}{P}$, there exists a subsequence $\underset{\sim}{P}_2$ of $\underset{\sim}{P}_1$ such that $A_t f_t \rightharpoonup g$ as $t \to \infty$ in $\underset{\sim}{P}_2$ for some $g \in \underset{\sim}{H}$. This is because bounded sets in $\underset{\sim}{H}$ are weakly conditionally compact. In particular if $h \in \underset{\sim}{C} = \{f: A_t f \to Af \text{ as } t \to \infty\}$

$$\langle A_t f_t | h\rangle \to \langle g|h\rangle \qquad \text{as } t \to \infty \text{ in } \underset{\sim}{P}_2 .$$

On the other hand

$$\langle A_t f_t | h\rangle = \langle f_t | A_t h\rangle$$

for all (large) t and thus

$$\langle A_t f_t | h\rangle \to \langle f | Ah\rangle \qquad \text{as } t \to \infty \text{ in } \underset{\sim}{P}_2 ,$$

so that

$$\langle g|h\rangle = \langle f|Ah\rangle .$$

Given $k \in \underset{\sim}{D}[A]$ and $\delta > 0$ there exists $h \in \underset{\sim}{C}$ such that $\|k-h\| < \delta$ and $\|Ak-Ah\| < \delta$. This implies that

$$\langle g|k\rangle = \langle f|Ak\rangle \qquad \text{for all } k \in \underset{\sim}{D}[A] .$$

Consequently $f \in \underset{\sim}{D}[A^*]$ and $A^*f = g$; but $A^* = A$. Since every subsequence $\underset{\sim}{P}_1$ contains a subsequence $\underset{\sim}{P}_2$ such that $A_t f_t \rightharpoonup Af$ as $t \to \infty$ in $\underset{\sim}{P}_2$, it follows that $A_t f_t \rightharpoonup Af$ as $t \to \infty$ in $\underset{\sim}{P}$.

In what follows in addition to i. and ii. of section 1 we assume that:

iii. $0 \leq S(t)$ is a bounded self-adjoint transformation on $\underset{\sim}{H}$, $t > 0$; $F(t)$ is a projection on $\underset{\sim}{H}$, $t > 0$;

iv. F is the strong limit of $F(t)$ as $t \to \infty$;

v. $S^{1/2}$ is the closure of the strong limit of $S(t)^{1/2}$ as $t \to \infty$;

vi. $S^{1/2}$ is the closure of the strong limit of $S(t)^{1/2}F(t)$ as $t \to \infty$ on $\underset{\sim}{M}$.

Let us set

$$S_F(t) = F(t)S(t)F(t) .$$

Theorem 2b. Under assumptions i.-vi. we have

$$\{S_F(t)-zI\}^{-1}f \rightharpoonup \{S_F-zI\}^{-1}f \qquad \text{as} \quad t \to \infty$$

for all $f \in \underset{\sim}{M}$ and all z for which $\text{Im } z \neq 0$.

Proof. Let $f \in \underset{\sim}{M}$. We will show that if $\underset{\sim}{P}_1$ is an arbitrary subsequence of $\underset{\sim}{P}$ then $\underset{\sim}{P}_1$ contains a subsequence $\underset{\sim}{P}_2$ such that

$$\{S_F(t)-zI\}^{-1}f \rightharpoonup \{S_F-zI\}^{-1}f$$

as $t \to \infty$ in $\underset{\sim}{P}_2$. This will prove our result. Since

$$\|\{S_F(t)-zI\}^{-1}f\| \leq |\text{Im } z|^{-1}\|f\| \qquad t \in \underset{\sim}{P}$$

we can find a subsequence $\underset{\sim}{P}_2$ of $\underset{\sim}{P}_1$ such that if $g(t) = \{S_F(t)-zI\}^{-1}f$ then $g(t) \rightharpoonup g$ as $t \to \infty$ in $\underset{\sim}{P}_2$ for some $g \in \underset{\sim}{H}$. We must show that $g = \{S_F-zI\}^{-1}f$. Since $F^* = F$ is the strong limit of $F(t)^* = F(t)$ we have

$$F(t)g(t) \rightharpoonup Fg , \qquad \text{as} \quad t \to \infty \quad \text{in } \underset{\sim}{P}_2 .$$

We also have

$$\langle S(t)^{1/2}F(t)g(t)|S(t)^{1/2}F(t)g(t)\rangle = \langle S_F(t)g(t)|g(t)\rangle$$

$$= \langle f+zg(t)|g(t)\rangle = O(1) .$$

Therefore by Lemma 2a we have $Fg \in \underset{\sim}{D}(S^{1/2})$ and

$$S(t)^{1/2}F(t)g(t) \rightharpoonup S^{1/2}Fg , \qquad t \to \infty \text{ in } \underset{\sim}{P}_2 .$$

We assert that $Fg = g$ which implies $g \in \underset{\sim}{S}$. Since $F(t)S_F(t) = S_F(t)$ we have

$$F(t)(f+zg(t)) = f + zg(t) .$$

Passing to the limit we get

$$F(f+zg) = f + zg .$$

But $Ff = f$ and $z \neq 0$ imply $Fg = g$.

Let $\underset{\sim}{S}' = \{f \in \underset{\sim}{M}: S(t)^{1/2}F(t)f \to S^{1/2}f \text{ as } t \to \infty\}$.

For $h \in \underset{\sim}{S}'$ it follows that

$$\lim_{\underset{\sim}{P}_2}\langle S_F(t)g(t)|h\rangle = \lim_{\underset{\sim}{P}_2}\langle S(t)^{1/2}F(t)g(t)|S(t)^{1/2}F(t)h\rangle$$

$$= \langle S^{1/2}g|S^{1/2}h\rangle .$$

On the other hand

$$\lim_{\underset{\sim}{P}_2}\langle S_F(t)g(t)|h\rangle = \lim_{\underset{\sim}{P}_2}\langle f+zg(t)|h\rangle$$

$$= \langle f+zg|h\rangle .$$

Thus

(1) $$\langle S^{1/2}g|S^{1/2}h\rangle = \langle f+zg|h\rangle \qquad \text{all } h \in \underset{\sim}{S}' .$$

By assumption vi. (1) holds for all $h \in \underset{\sim}{S}$, and thus for all $h \in \underset{\sim}{D}(S_F)$. For such an h we have, by (2) of § 1,

$$\langle S^{1/2}g | S^{1/2}h \rangle = \langle g | S_F h \rangle .$$

Consequently $\langle f+zg | h \rangle = \langle g | S_F h \rangle$ or equivalently

$$\langle g | \{S_F - z^*I\} h \rangle = \langle f | h \rangle$$

for all $h \in \underset{\sim}{D}(S_F)$. This implies that $g \in \underset{\sim}{D}((S_F - z^*I)^*)$,

$\{S_F - z^*I\}^* g = f$, and hence $g = \{S_F - zI\}^{-1} f$.

3. Spectral Resolutions

Let

$$S_F(t) = \int_{0-}^{\infty} \lambda \, d\Psi(t,\lambda)$$

be the spectral resolution of $S_F(t)$ on $\underset{\sim}{H}$ and

$$S_F = \int_{0-}^{\infty} \lambda \, d\Psi(\lambda)$$

be the spectral resolution of S_F on $\underset{\sim}{M}$. We assume throughout that $\Psi(t,\lambda) = \Psi(t,\lambda+)$, $0 \leq \lambda < \infty$, $t > 0$, that $\Psi(t,0-) = 0$, and similarly for $\Psi(\lambda)$.

Theorem 3a. If λ is not in $\sigma_p(S_F)$, the point spectrum of S_F, then

$$\Psi(t,\lambda) f \to \Psi(\lambda) f \qquad \text{as} \quad t \to \infty$$

for all $f \in \underset{\sim}{M}$.

Proof. By Theorem 2b if $\operatorname{Im} z \neq 0$ then

$$\lim_{t \to \infty} \int_{0-}^{\infty} (\lambda - z)^{-1} d_\lambda \langle \Psi(t,\lambda) f | f \rangle = \int_{0-}^{\infty} (\lambda - z)^{-1} d_\lambda \langle \Psi(\lambda) f | f \rangle .$$

The set of functions $(\lambda - z)^{-1}$, $\text{Im } z \neq 0$, is fundamental in $C_o(\underline{R})$, the space of all complex valued continuous functions on $\underline{R}$ vanishing at ∞ , taken with the uniform norm. It follows that

$$(1) \qquad \lim_{t \to \infty} \int \varphi(\lambda) d_\lambda \langle \Psi(t,\lambda) f | f \rangle = \int \varphi(\lambda) d_\lambda \langle \Psi(\lambda) f | f \rangle$$

for all $\varphi(\lambda) \in C_o(\underline{R})$. We note that $\langle \Psi(\lambda) f | f \rangle$ and $\langle \Psi(t,\lambda) f | f \rangle$ are non-decreasing functions of λ . The application of (1) to suitably chosen trapezoidal functions, yields

$$\overline{\lim_{t \to \infty}} \ \langle \Psi(t,\lambda) f | f \rangle \leq \langle \Psi(\lambda+\varepsilon) f | f \rangle ,$$

and

$$\underline{\lim_{t \to \infty}} \ \langle \Psi(t,\lambda) f | f \rangle \geq \langle \Psi(\lambda-\varepsilon) f | f \rangle ,$$

for any λ and all $\varepsilon > 0$. If λ is not in the point spectrum of S_F then, from the continuity of Ψ at λ , we obtain

$$\lim_{t \to \infty} \langle \Psi(t,\lambda) f | f \rangle = \langle \Psi(\lambda) f | f \rangle .$$

Using the polarization identity

$$\langle Ef | g \rangle = \tfrac{1}{4}\|E(f+g)\|^2 - \tfrac{1}{4}\|E(f-g)\|^2 + \tfrac{i}{4}\|E(f+ig)\|^2 - \tfrac{i}{4}\|E(f-ig)\|^2 ,$$

valid for any projection E in $\underset{\sim}{H}$, we see that

$$\lim_{t \to \infty} \langle \Psi(t,\lambda) f | g \rangle = \langle \Psi(\lambda) f | g \rangle \qquad \text{all } f,g \in \underset{\sim}{M} ,$$

provided λ is not in the point spectrum of S_F . It is a well known elementary fact that weak convergence for projections implies strong convergence. Our proof is now complete.

Theorem 3b. If λ is not in $\sigma_p(S_F)$ then

$$\Psi(t,\lambda)g \to 0 \qquad \text{as} \quad t \to \infty$$

for all $g \in \underset{\sim}{M}^{\perp} \cap \underset{\sim}{H}$.

Proof. Let $g \in \underset{\sim}{M}^{\perp} \cap \underset{\sim}{H}$. Since $\|\Psi(t,\lambda)g\| = O(1)$, given any subsequence $\underset{\sim}{P}_1$ there is a subsequence $\underset{\sim}{P}_2$ of $\underset{\sim}{P}_1$ such that

$$\Psi(t,\lambda)g \rightharpoonup h \qquad \text{as} \quad t \to \infty \quad \text{in } \underset{\sim}{P}_2$$

for some $h \in \underset{\sim}{H}$. Thus if $f \in \underset{\sim}{M}$

$$\lim_{\underset{\sim}{P}_2} \langle \Psi(t,\lambda)g | f \rangle = \langle h | f \rangle .$$

Since $\langle \Psi(t,\lambda)g | f \rangle = \langle g | \Psi(t,\lambda)f \rangle$ we have by Theorem 3a

$$\lim_{\underset{\sim}{P}_2} \langle \Psi(t,\lambda)g | f \rangle = \langle g | \Psi(\lambda)f \rangle = 0 .$$

Thus $\langle h | f \rangle = 0$; i.e. $h \perp \underset{\sim}{M}$. Since

$$F(t)S_F(t) = S_F(t)F(t) = S_F(t)$$

we have

$$F(t)\Psi(t,\lambda) = \Psi(t,\lambda)F(t) .$$

Using $g = Fg$ we see that

$$\begin{aligned} \Psi(t,\lambda)g &= \Psi(t,\lambda)F(t)g + \Psi(t,\lambda)(F-F(t))g \\ &= F(t)\Psi(t,\lambda)g + o(1) \quad \text{as} \quad t \to \infty . \end{aligned}$$

Passing to the limit we find that $h = Fh$ so that $h \in F\underset{\sim}{H}$. We have

$$\begin{aligned} \|S(t)^{1/2}F(t)\Psi(t,\lambda)g\|^2 &= \langle S_F(t)\Psi(t,\lambda)g | g \rangle \\ &= \int_{0-}^{\lambda} \mu \, d_\mu \|\Psi(t,\mu)g\|^2 \end{aligned}$$

$$\leq \lambda \|g\|^2 .$$

Therefore by Lemma 2a $h \in \underset{\sim}{D}[S^{1/2}]$; i.e. $h \in \underset{\sim}{S} \subset \underset{\sim}{M}$. But $h \perp \underset{\sim}{M}$ so that $h = 0$.

These results can be put into a more convenient form. Let

$$S_F(t)\Big|_{F(t)\underset{\sim}{H}} = \int_{0-}^{\infty} \lambda d\Phi(t,\lambda)$$

be the spectral resolution of $S_F(t)$ restricted to $F(t)\underset{\sim}{H}$. It is not hard to see that

$$\Phi(t,\lambda) = \Psi(t,\lambda)F(t) .$$

Because of this formula each $\Phi(t,\lambda)$ can be regarded as a projection on $\underset{\sim}{H}$. However, $\Phi(t,\lambda)$, $-\infty < \lambda < \infty$, is a resolution of the identity only on $F(t)\underset{\sim}{H}$. The operator S_F is defined only on $\underset{\sim}{M}$ and the $\Psi(\lambda)$ are projections on $\underset{\sim}{M}$. Let us set

$$\Phi(\lambda) = \Psi(\lambda)G$$

where G is the projection of $\underset{\sim}{H}$ onto $\underset{\sim}{M}$. $\Phi(\lambda)\big|_{\underset{\sim}{M}} \equiv \Psi(\lambda)$, but $\Phi(\lambda)$, unlike $\Psi(\lambda)$, is defined on $\underset{\sim}{H}$.

Theorem 3c. If assumptions i - vi hold then for $\lambda \notin \sigma_p(S_F)$

$$\Phi(t,\lambda) \to \Phi(\lambda) \qquad \text{as} \quad t \to +\infty .$$

Here " $\to$ " is strong convergence of operators on $\underset{\sim}{H}$.

Proof. We have

$$\Phi(t,\lambda) = \Psi(t,\lambda)[F(t)-F] + \Psi(t,\lambda)F .$$

Because $F(t) \to F$ as $t \to \infty$ we see that

$$\Phi(t,\lambda) - \Psi(t,\lambda)F \to 0 \qquad \text{as} \quad t \to +\infty .$$

In conjunction with Theorems 3a and 3b this implies Theorem 3c.

4. Convergence in Dimension

Theorem 4a stated below shows how starting from the conclusion of Theorem 3c and one additional assumption it is possible to prove that the dimensions of the spectral projections converge. See Widom [14], and also [3].

Suppose that $0 \leq R(t)$, $t > 0$ are bounded positive self-adjoint operators defined on subspaces $\underset{\sim}{N}(t)$ of a Hilbert space $\underset{\sim}{H}$. Let $0 \leq R$ be a positive self-adjoint operator on a subspace $\underset{\sim}{M}$ of $\underset{\sim}{H}$. Let

$$R(t) = \int_{0-}^{\infty} \lambda dE(t,\lambda) ,$$

$$R = \int_{0-}^{\infty} \lambda dE(\lambda) ,$$

be the spectral resolutions of $R(t)$ on $\underset{\sim}{N}(t)$ and of R on $\underset{\sim}{M}$. Let us regard $E(t,\lambda)$ as the projection of $\underset{\sim}{H}$ onto $E(t,\lambda)\underset{\sim}{N}(t)$ and $E(\lambda)$ as the projection of $\underset{\sim}{H}$ onto $E(\lambda)\underset{\sim}{M}$. This usage is consistent with the definitions of $E(t,\lambda)$ and $E(\lambda)$. Note, however, that $E(t,\lambda)$ and $E(\lambda)$ are not spectral resolutions on $\underset{\sim}{H}$. Consider the following assumptions:

a. for $\lambda \notin \sigma_p(R)$, $E(t,\lambda) \to E(\lambda)$ in $\underset{\sim}{H}$ as $t \to +\infty$;

b. there is a number $m > 0$ such that if $f_t \in \underset{\sim}{N}(t)$, $\|f_t\| = 1$, and if $\langle R(t)f_t | f_t \rangle \leq m_1 < m$ for $t \in \underset{\sim}{P}_1$, then $\underset{\sim}{P}_1$ contains a subsequence $\underset{\sim}{P}_2$ such that $f_t \rightharpoonup f \neq 0$ as $t \to \infty$ in $\underset{\sim}{P}_2$.

Theorem 4a. Under assumptions a. and b. we have

(1) $$\dim E(\lambda) < \infty , \qquad 0 \leq \lambda < m ,$$

and

(2) $$\lim_{t \to \infty} \dim E(t,\lambda) = \dim E(\lambda)$$

for $0 \leq \lambda < m$, $\lambda \notin \sigma_p(R)$.

Proof. We first assert that assumption a. above implies that if $0 \leq \lambda < \infty$, $\lambda \notin \sigma_p(R)$, then

(3) $$\underline{\lim}_{t \to \infty} \dim E(t,\lambda) \geq \dim E(\lambda) .$$

In (3) we admit $\infty \geq \infty$. Suppose $\dim E(\lambda) \geq k$. Then there exist orthonormal vectors $g_1, g_2, \ldots, g_k$ in $E(\lambda)\underset{\sim}{H}$. By assumption a. we have

$$\lim_{t \to \infty} E(t,\lambda)g_j = E(\lambda)g_j = g_j \qquad j = 1, \ldots, k$$

from which it follows that for all sufficiently large t $\{E(t,\lambda)g_j\}_1^k$, which belong to $E(t,\lambda)\underset{\sim}{N}(t)$, are linearly independent, so that $\dim E(t,\lambda)\underset{\sim}{N}(t) \geq k$.

From this point on we use assumptions a. and b. We suppose $\lambda \notin \sigma_p(R)$ and that $0 \leq \lambda < m$. If $\dim E(\lambda) = \infty$ then we can find an infinite orthonormal set of vectors $\{g_k\}_1^\infty$ in $E(\lambda)\underset{\sim}{H}$. Using a. we see that there exists a subsequence $\underset{\sim}{P}_1 = \{0 < t_1 < t_2 < \ldots\}$ such that $\|E(t_k,\lambda)g_k - g_k\| \to 0$ as $k \to \infty$. Let us set $f_k = E(t_k,\lambda)g_k / \|E(t_k,\lambda)g_k\|$. Note that $E(t_k,\lambda)g_k \neq 0$ for k "large". For such k $\|f_k\| = 1$, and $\langle R(t_k)f_k | f_k \rangle \leq \lambda$ $k = 1, 2, \ldots$

Therefore by b. there is a subsequence $\underset{\sim}{P}_2$ of $\{1, 2, \ldots\}$ such that $f_k \rightharpoonup f \neq 0$ as $k \to \infty$ in $\underset{\sim}{P}_2$. But then $g_k \rightharpoonup f \neq 0$ as $k \to \infty$ in $\underset{\sim}{P}_2$. However it is obvious that $g_n \rightharpoonup 0$ as $n \to \infty$. Thus $\dim E(\lambda) = \infty$ leads to a contradiction and (1) is true.

We assert that (2) is true. Set $k = \dim E(\lambda)$. If (2) is not true then in view of (3) there is a subsequence $\underset{\sim}{P}_1$ such that $\dim E(t,\lambda)\underset{\sim}{N} > k$ for $t \in \underset{\sim}{P}_1$. Let $g_1, \ldots, g_k$ be an orthonormal basis for $E(\lambda)\underset{\sim}{N}$. For each $t \in \underset{\sim}{P}_1$ we can choose $f_t \in E(t,\lambda)\underset{\sim}{H}$ such that $\|f_t\| = 1$, $f_t \perp g_1, \ldots, g_k$. We have $\langle R(t)f_t | f_t\rangle \leq \lambda$ and therefore by b. there is a subsequence $\underset{\sim}{P}_2$ of $\underset{\sim}{P}_1$ such that $f_t \rightharpoonup f \neq 0$ as $t \to \infty$ in $\underset{\sim}{P}_2$. Now $f_t = E(t,\lambda)f_t$ and by a. $E(t,\lambda)f_t \rightharpoonup E(\lambda)f$ as $t \to \infty$ in $\underset{\sim}{P}_2$. Therefore $f = E(\lambda)f$ and $f \in E(\lambda)\underset{\sim}{N}$. Since $f \perp g_1, \ldots, g_k$ f must be 0 . This is a contradiction and our assertion follows.

Chapter III

THE FOURIER TRANSFORM THEOREM

1. Spaces and Operators

We begin by listing some assumptions which will be in force throughout sections 1 - 7. In section 8 iv. and v. will be replaced by more general conditions.

i. We have a set Ω in $\underline{R}_n$ which is of finite positive measure and is locally star-shaped. We recall that a set Ω in $\underline{R}_n$ is star-shaped with respect to the origin if for each $r < 1$ the closure of $r\Omega$ is contained in the interior of Ω. Ω is star-shaped if some translate of Ω is star-shaped with respect to the origin. Ω is locally star-shaped if every point of $\bar{\Omega}$ (the closure of Ω) has a neighborhood whose intersection Ω is star-shaped.

ii. We have a real function $f^{\wedge}(\underline{\xi})$ on $\underline{R}_n$ which is bounded and measurable, which assumes its minimum 0 at precisely p distinct points $\underline{\xi}_m$ $m = 1, \ldots, p$, and is such that

$$\inf \{ f^{\wedge}(\underline{\xi}) : \ |\underline{\xi}-\underline{\xi}_m| \geq \delta \quad m = 1, \ldots, p \} > 0$$

for each $\delta > 0$.

iii. We are given a positive constant ω and a positive continuous function $L(t)$ defined for $0 < t < \infty$ which is slowly oscillating at 0. $L(t)$ is slowly oscillating at 0 if for every $\epsilon > 0$ there is an $\alpha(\epsilon) > 0$ such that $L(t)t^{\epsilon}$ is increasing and $L(t)t^{-\epsilon}$ is decreasing for $0 < t < \alpha(\epsilon)$.

iv. We have p non-negative measurable functions $\Phi_m(\underline{\xi})$ which are homogeneous of degree 0 on $\underline{R}_n$ and which are bounded and bounded away from 0 .

v. For each $m = 1, \ldots, p$ given $\varepsilon > 0$ there exists a $\delta > 0$ such that $0 < |\underline{\xi}-\underline{\xi}_m| < \delta$ implies that

$$\left| \frac{f^\wedge(\underline{\xi})}{L(|\underline{\xi}-\underline{\xi}_m|)|\underline{\xi}-\underline{\xi}_m|^{\omega}} - \Phi_m(\underline{\xi}-\underline{\xi}_m) \right| < \varepsilon .$$

It will be necessary to work with four Hilbert spaces which we will denote by $\underset{\sim}{L}$, $\underset{\sim}{L}^\wedge$, $\underset{\sim}{H}$ and $\underset{\sim}{H}^\wedge$.

Let L be the Hilbert space of complex Lebesgue measurable functions $u(\underline{\eta}), v(\underline{\eta})$, etc. on $\underline{R}_n$ defined by the inner product

$$\langle u | v \rangle_{\underset{\sim}{L}} = \int_{\underline{R}_n} u(\underline{\eta})\overline{v(\underline{\eta})}d\underline{\eta} .$$

Here $\underline{\eta} = (\eta_1, \ldots, \eta_n)$ and $d\underline{\eta}$ is Lebesgue measure on $\underline{R}_n$. We set $\|u\|_{\underset{\sim}{L}} = \langle u | u \rangle_{\underset{\sim}{L}}^{1/2}$. Let $\underset{\sim}{L}^\wedge$ be the Hilbert space of complex measurable functions $u^\wedge(\underline{\xi})$, $v^\wedge(\underline{\xi})$, etc. on $\underline{R}_n$ defined by the inner product

$$\langle u^\wedge | v^\wedge \rangle_{L^\wedge} = (2\pi)^{-n} \int_{\underline{R}_n} u^\wedge(\underline{\xi}) v^\wedge(\underline{\xi}) d\underline{\xi} .$$

We set

$$\|u^\wedge\|_{\underset{\sim}{L}^\wedge} = \langle u^\wedge | u^\wedge \rangle_{\underset{\sim}{L}^\wedge}^{1/2} .$$

The mapping φ from $\underset{\sim}{L}$ to $\underset{\sim}{L}^\wedge$ (formally) defined by

$$u^\wedge(\underline{\xi}) = \varphi u \cdot (\underline{\xi}) = \int_{\underline{R}_n} u(\underline{\eta}) e^{i\underline{\xi}\cdot\underline{\eta}} d\underline{\eta} .$$

is unitary. It is inverse φ^{-1} is given by

$$\varphi^{-1}u^{\wedge}\cdot(\underline{\eta}) = (2\pi)^{-n}\int_{\underline{R}_n} u^{\wedge}(\underline{\xi})e^{-i\underline{\xi}\cdot\underline{\eta}}d\underline{\xi}$$

Here $\underline{\xi}\cdot\underline{\eta} = \xi_1\eta_1 + \dots + \xi_n\eta_n$.

Let Ω , ω , $f(\underline{\xi})$, etc. be as in assumptions i. - v. Let $0 < t < \infty$. We define $E(t)$ on $\underset{\sim}{L}$ by

$$E(t)u\cdot(\underline{\eta}) = \begin{cases} u(\underline{\eta}) & \underline{\eta} \in t\Omega \\ 0 & \underline{\eta} \notin t\Omega \end{cases} \qquad u \in L .$$

If $E^{\wedge}(t)$ on $\underset{\sim}{L}^{\wedge}$ is defined by

$$E^{\wedge}(t) = \varphi E(t)\varphi^{-1} ,$$

then

$$E^{\wedge}(t)u^{\wedge}\cdot(\underline{\xi}) = \int_{t\Omega} (\varphi^{-1}u^{\wedge})\cdot(\underline{\eta})e^{i\underline{\xi}\cdot\underline{\eta}}d\underline{\eta} .$$

It is evident that $E(t)$ is a projection on $\underset{\sim}{L}$. $E^{\wedge}(t)$ which is unitarily equivalent to $E(t)$ is therefore a projection on $\underset{\sim}{L}^{\wedge}$. We define $T^{\wedge}$ on $\underset{\sim}{L}^{\wedge}$ by

$$T^{\wedge}u^{\wedge}\cdot(\underline{\xi}) = f(\underline{\xi})\ u^{\wedge}(\underline{\xi}) .$$

Since $f(\underline{\xi})$ is non-negative and bounded $T^{\wedge}$ is a non-negative, bounded, self-adjoint transformation. It follows that if we set

$$T_E^{\wedge}(t) = E^{\wedge}(t)T^{\wedge}E^{\wedge}(t)\Big|_{E^{\wedge}(t)\underset{\sim}{L}^{\wedge}} ,$$

where the right hand side is to read $E^{\wedge}(t)T^{\wedge}E^{\wedge}(t)$ restricted to (the Hilbert space) $E^{\wedge}(t)\underset{\sim}{L}^{\wedge}$, then $T_E^{\wedge}(t)$ is a strictly positive bounded self-adjoint operator. It is to the study of $T_E^{\wedge}(t)$ that the present chapter is devoted.

We will show that as $t \to +\infty$ "the bottom of the spectrum of $T_E^{\wedge}(t)$ becomes discretized"; that is, given any positive integer N there exists $t(N)$ with the following property: for each $t > t(N)$ there is a value $\eta(t,N) > 0$ such that the spectrum of $T_E^{\wedge}(t)$ in $[0,\eta(t,N)]$ consists of a finite number of eigen values, each of finite multiplicity and the number of these eigen values (taking into account their multiplicities) is at least N. Thus if $t > t(N)$ it makes sense to speak of the lowest N-eigen values of $T_E^{\wedge}(t)$ repeated according to their multiplicities, arranged in non-decreasing order,

$$0 < \lambda(1,t) \leq \lambda(2,t) \leq \cdots \leq \lambda(N,t) .$$

We will further show that for each $m = 1, \ldots, p$ there exists a sequence $\{\mu_m(k)\}_1^\infty$

$$0 < \mu_m(1) \leq \mu_m(2) \leq \ldots, \quad \lim_{k \to \infty} \mu_m(k) = \infty ,$$

depending only on the triple,

$$(\Omega, \omega, \Phi_m) ,$$

such that if $\{\mu(k)\}_1^\infty$ is the set $\bigcup_{m=1}^{p} \{\mu_m(k)\}_1^\infty$ arranged in non-decreasing order, then

$$\lim_{t \to \infty} t^{\omega} L(t^{-1})^{-1} \lambda(k,t) = \mu(k) \qquad k = 1, 2, \ldots .$$

We define $\underset{\sim}{H}$ and $\underset{\sim}{H}^{\wedge}$ to be Hilbert spaces of complex measurable functions defined in $\underline{R}_n$ with inner products

$$\langle u|v\rangle_{\underset{\sim}{H}} = \int_{\underline{R}_n} u(\underline{y})\overline{v(\underline{y})}d\underline{y} \ ,$$

$$\langle u^{\wedge}|v^{\wedge}\rangle_{\underset{\sim}{H}^{\wedge}} = (2\pi)^{-n} \int_{\underline{R}_n} u^{\wedge}(\underline{x})\overline{v^{\wedge}(\underline{x})}d\underline{x} \ .$$

The mapping ψ from $\underset{\sim}{H}$ to $\underset{\sim}{H}^{\wedge}$ defined by

$$\psi u\cdot(\underline{x}) = \int_{\underline{R}_n} u(\underline{y})e^{i\underline{x}\cdot\underline{y}}d\underline{y}$$

and its inverse ψ^{-1} from $\underset{\sim}{H}^{\wedge}$ to $\underset{\sim}{H}$ defined by

$$\psi^{-1}u^{\wedge}\cdot(\underline{y}) = (2\pi)^{-n} \int_{\underline{R}_n} u^{\wedge}(\underline{x})e^{-i\underline{y}\cdot\underline{x}}d\underline{x}$$

are unitary mappings. As indicated in the above definitions we will reserve the variables $\underline{\eta}$, $\underline{\xi}$, $\underline{y}$, $\underline{x}$ for elements of $\underset{\sim}{L}$, $\underset{\sim}{L}^{\wedge}$, $\underset{\sim}{H}$, $\underset{\sim}{H}^{\wedge}$, respectively.

$\underset{\approx}{H}$ and $\underset{\approx}{H}^{\wedge}$ consist of elements which are ordered p-tuples of functions in $\underset{\sim}{H}$ and $\underset{\sim}{H}^{\wedge}$ respectively, i.e. if $\underline{u}(\underline{y}) \in \underset{\approx}{H}$ then

$$\underline{u}(\underline{y}) = [u_1(\underline{y}), \ldots, u_p(\underline{y})] \ , \quad u_j \in \underset{\sim}{H} \ , \quad 1 \leq j \leq p$$

and if $\underline{u}^{\wedge}(\underline{x}) \in \underset{\approx}{H}^{\wedge}$ then

$$\underline{u}^{\wedge}(\underline{x}) = [u_1^{\wedge}(\underline{x}), \ldots, u_p^{\wedge}(\underline{x})] \ , \quad u_j^{\wedge} \in \underset{\sim}{H}^{\wedge} \ , \quad 1 \leq j \leq p \ .$$

The inner product in $\underset{\approx}{H}$ is defined by

$$\langle \underline{u}|\underline{v}\rangle_{\underset{\approx}{H}} = \langle [u_1, \ldots, u_p]|[v_1, \ldots, v_p]\rangle_{H}$$

$$= \sum_{m=1}^{p} \langle u_m|v_m\rangle_{\underset{\sim}{H}} \ .$$

Similarly if $\underline{u}^{\wedge}$, $\underline{v}^{\wedge} \in \underset{\approx}{H}^{\wedge}$

$$\langle \underline{u}^\wedge | \underline{v}^\wedge \rangle_{\underset{\sim}{H}^\wedge} = \langle [u_1^\wedge, \ldots, u_p^\wedge] | [v_1^\wedge, \ldots, v_p^\wedge] \rangle_{\underset{\sim}{H}^\wedge}$$

$$= \sum_{m=1}^{p} \langle u_m^\wedge | v^\wedge_m \rangle_{\underset{\sim}{H}^\wedge} .$$

The mapping Ψ from $\underset{\sim}{H}$ to $\underset{\sim}{H}^\wedge$ defined by

$$\Psi \underline{u} \cdot (\underline{x}) = \Psi [u_1, \ldots, u_p] \cdot (\underline{x}) = [\psi u_1 \cdot (\underline{x}), \ldots, \psi u_p \cdot (\underline{x})]$$

and its inverse Ψ^{-1} from $\underset{\sim}{H}^\wedge$ to $\underset{\sim}{H}$, defined by

$$\Psi^{-1} \underline{u}^\wedge \cdot (\underline{y}) = \Psi^{-1} [u_1^\wedge, \ldots, u_p^\wedge] \cdot (\underline{y}) = [\psi^{-1} u_1^\wedge \cdot (\underline{y}), \ldots, \psi^{-1} u_p^\wedge \cdot (\underline{y})] ,$$

are unitary mappings.

Let $\underline{a} = (a_1, \ldots, a_n)$, $\underline{b} = (b_1, \ldots, b_n)$ be extended real-valued vectors such that $-\infty \le a_j$ and $b_j \le +\infty$, $j = 1, \ldots, n$. By a rectangle $[\underline{a}, \underline{b}]$ we mean the set

$$\{\underline{z} = (z_1, \ldots, z_n) \in \underline{R}_n : a_j \le z_j \le b_j , 1 \le j \le n\} .$$

Clearly $\underline{R}_n$ can be represented as the union of p non-overlapping rectangles each of which contains exactly one of the points $\underline{\xi}_m$, $m = 1, \ldots, p$ in its interior. We denote this partition by $R_1 = [\underline{a}_1, \underline{b}_1]$, ..., $R_p = [\underline{a}_p, \underline{b}_p]$.

For each $t > 0$, $m = 1, \ldots, p$ we define a mapping $\sigma(t,m)$ of R_m into $\underline{R}_n$ by

$$\underline{x} = \sigma(t,m) \cdot (\underline{\xi}) = t(\underline{\xi} - \underline{\xi}_m) .$$

We denote by R_m^t the image of R_m under this mapping. Similarly the inverse of $\sigma(t,m)$, denoted $\tau(t,m)$ is defined by

$$\underline{\xi} = \tau(t,m)\cdot(\underline{x}) = t^{-1}\underline{x} + \underline{\xi}_m .$$

For each $t > 0$, $m = 1, \ldots, p$ we define a mapping $\chi(t,m)$ from $\underset{\sim}{L}^{\wedge}$ to $\underset{\sim}{H}^{\wedge}$ as follows. If $u^{\wedge}(\underline{\xi}) \in \underset{\sim}{L}^{\wedge}$

$$\chi(t,m)u^{\wedge}\cdot(\underline{x}) = \begin{cases} t^{-n/2}u^{\wedge}\cdot\tau(t,m)(\underline{x}) & , \underline{x} \in R^t_m \\ 0 & , \underline{x} \notin R^t_m . \end{cases}$$

Next we define a mapping $\underline{\chi}(t)$ from $\underset{\sim}{L}^{\wedge}$ to $\underset{\approx}{H}^{\wedge}$ by

$$\underline{\chi}(t)u^{\wedge}(\underline{x}) = [\chi(t,1)u^{\wedge}(\underline{x}), \ldots, \chi(t,p)u^{\wedge}(\underline{x})] .$$

We claim that $\underline{\chi}(t)$, $t > 0$ is an isometry. We have

$$\text{(1)} \qquad \|u^{\wedge}\|^2_{\underset{\sim}{L}^{\wedge}} = (2\pi)^{-n} \int_{\underline{R}_n} |u^{\wedge}(\underline{\xi})|^2 d\underline{\xi} = \sum_{m=1}^{p} \int_{R_m} |u^{\wedge}(\underline{\xi})|^2 d\underline{\xi} .$$

On the other hand

$$\|\chi(t,m)u^{\wedge}\|^2_{\underset{\sim}{H}^{\wedge}} = (2\pi)^{-n} \int_{R^t_m} t^{-n} |u^{\wedge}(t^{-1}\underline{x}+\underline{\xi}_m)|^2 d\underline{x}.$$

By the change of variables $\underline{\xi} = t^{-1}\underline{x} + \underline{\xi}_m$ this last expression becomes

$$(2\pi)^{-n} \int_{R_m} |u^{\wedge}(\underline{\xi})|^2 d\underline{\xi} .$$

Therefore

$$\text{(2)} \qquad \|\underline{\chi}(t)u^{\wedge}\|^2_{\underset{\approx}{H}^{\wedge}} = \int_{R_m} |u^{\wedge}(\underline{\xi})|^2 d\underline{\xi} .$$

(1) and (2) prove our claim.

The adjoint of $\underline{\chi}(t)$ is defined as follows. We first define

$\chi^*(t,m)$ mapping $\underset{\sim}{H}^\wedge$ into $\underset{\sim}{L}^\wedge$ for each $t > 0$, $m = 1, \ldots, p$ by

$$\chi^*(t,m)u^\wedge\cdot(\underline{\xi}) = \begin{cases} t^{n/2}u^\wedge\cdot\sigma(t,m)(\underline{\xi}) , & \underline{\xi} \in R_m \\ 0 , & \underline{\xi} \notin R_m \end{cases} .$$

If $\underline{u}^\wedge = [u_1^\wedge, \ldots, u_p^\wedge] \in \underset{\approx}{H}^\wedge$ then $\underline{\chi}^*(t)\underline{u}^\wedge(\underline{\xi})$, $\underline{\xi} \in \underline{R}_n$ is defined by

$$\underline{\chi}^*(t)\underline{u}^\wedge\cdot(\underline{\xi}) = \begin{cases} \chi^*(t,1)u_1^\wedge(\underline{\xi}) , & \underline{\xi} \in R_1 \\ \vdots & \vdots \\ \chi^*(t,p)u_p^\wedge(\underline{\xi}) , & \underline{\xi} \in R_p \end{cases} .$$

This function is well defined at each $\underline{\xi}$ since the R_m's are a partition of $\underline{R}_n$ into disjoint subsets. We now verify that $\underline{\chi}^*(t)$ is in fact the adjoint of $\underline{\chi}(t)$. We must check that

$$\langle \underline{\chi}(t)u^\wedge | \underline{v}^\wedge \rangle_{\underset{\approx}{H}^\wedge} = \langle u^\wedge | \underline{\chi}^*(t)\underline{v}^\wedge \rangle_{\underset{\sim}{L}^\wedge}$$

for all $u^\wedge \in \underset{\sim}{L}^\wedge$, $\underline{v}^\wedge \in \underset{\approx}{H}^\wedge$.

$$\langle \underline{\chi}(t)u^\wedge | \underline{v}^\wedge \rangle_{\underset{\approx}{H}^\wedge} = \sum_{m=1}^{p} \langle \chi(t,m)u^\wedge | v_m^\wedge \rangle_{\underset{\sim}{H}^\wedge}$$

$$= (2\pi)^{-n} \sum_{m=1}^{p} \int_{R_m^t} t^{-n/2} u^\wedge\cdot\tau(t,m)(\underline{x})\overline{v_m^\wedge(\underline{x})}d\underline{x}$$

$$= (2\pi)^{-n} \sum_{m=1}^{p} \int_{R_m^t} t^{-n/2} u^\wedge(t^{-1}\underline{x} + \underline{\xi}_m)\overline{v_m^\wedge(\underline{x})}dx$$

$$= (2\pi)^{-n} \sum_{m=1}^{p} \int_{R_m} u^\wedge(\underline{\xi})t^{n/2}\, \overline{v_m^\wedge\cdot\sigma(t,m)(\underline{\xi})}d\underline{\xi}$$

where we have made the change of variables $\underline{\xi} = t^{-1}\underline{x} + \underline{\xi}_m$. This last sum clearly is equal to

$$(2\pi)^{-n} \sum_{m=1}^{p} \int_{R_m} u^{\wedge}(\underline{\xi})\overline{X^*(t,m)v_m^{\wedge}(\underline{\xi})}\,d\underline{\xi}$$

$$= \langle u^{\wedge} | \underline{X}^*(t)\underline{v}^{\wedge}\rangle_{\underset{\sim}{L}^{\wedge}} .$$

We have

$$\underline{X}^*(t)\underline{X}(t) = I$$

$$\underline{X}(t)\underline{X}^*(t) = \begin{cases} I & \text{on } \underline{X}(t)\underset{\sim}{L}^{\wedge} \\ 0 & \text{on } (\underline{X}(t)\underset{\sim}{L}^{\wedge})^{\perp} \end{cases} .$$

2. The Application of the Perturbation Theory

In this section we introduce various operators on the Hilbert spaces defined in section 1.

a) We recall that $E(t)$ is defined on $\underset{\sim}{L}$ by the formula

$$E(t)u(\underline{\eta}) = \begin{cases} u(\underline{\eta}) & \underline{\eta} \in t\Omega \\ 0 & \underline{\eta} \notin t\Omega \end{cases} ,$$

and that $E^{\wedge}(t)$ on $\underset{\sim}{L}^{\wedge}$ is defined by $E^{\wedge}(t) = \phi E(t)\phi^{-1}$. We then have

$$E^{\wedge}(t)u^{\wedge}(\underline{\xi}) = \int_{t\Omega} (\phi^{-1}u^{\wedge})(\underline{\eta})e^{i\underline{\xi}\cdot\underline{\eta}}d\underline{\eta} .$$

$\underline{F}^{\wedge}(t)$ on $\underset{\approx}{H}^{\wedge}$ is defined by $\underline{F}^{\wedge}(t) = \underline{X}(t)E^{\wedge}(t)\underline{X}^*(t)$. We note that

$E(t)$ is a projection on $\underset{\sim}{L}$,

$E^{\wedge}(t)$ is a projection on $\underset{\sim}{L}^{\wedge}$,

$\underline{F}^{\wedge}(t)$ is a projection on $\underset{\sim}{H}^{\wedge}$.

For each $t > 0$ we also define p^2 operators on $\underset{\sim}{H}^{\wedge}$, $F^{\wedge}(t,\ell,m)$ $\ell,m = 1, \ldots, p$ by

$$F^{\wedge}(t,\ell,m)u^{\wedge} = \chi(t,\ell)E^{\wedge}(t)\chi^*(t,m)u^{\wedge} .$$

We note that each of these operators is bounded with norm equal to one and if $\ell = m$ the operator is in fact a projection. Let us examine the action of $F^{\wedge}(t,\ell,m)$. Let $u^{\wedge} \in \underset{\sim}{H}^{\wedge}$ then

$$\chi^*(t,m)u^{\wedge}(\underline{\xi}) = \begin{cases} t^{n/2}u^{\wedge}(t(\underline{\xi}-\underline{\xi}_m)) , & \underline{\xi} \in R_m \\ 0 , & \underline{\xi} \notin R_m \end{cases} .$$

We have

$$\phi^{-1}(\chi^*(t,m)u)^{\wedge}(\underline{\eta}) = (2\pi)^{-n} \int_{\underline{R}_n} \chi^*(t,m)u^{\wedge}(\underline{\xi})e^{-i\underline{\xi}\cdot\underline{\eta}}d\underline{\xi}$$

$$= (2\pi)^{-n} \int_{R_m} t^{n/2}u^{\wedge}(t(\underline{\xi}-\underline{\xi}_m))e^{-i\underline{\xi}\cdot\underline{\eta}}d\underline{\xi}$$

$$= (2\pi)^{-n} \int_{R_m^t} t^{-n/2}u^{\wedge}(\underline{z})e^{-i(t^{-1}\underline{z}+\underline{\xi}_m)\cdot\underline{\eta}}d\underline{z}$$

where we have made the change of variables $t(\underline{\xi}-\underline{\xi}_m) = \underline{z}$. Therefore

$(E^{\wedge}(t)\chi^*(t,m)u^{\wedge})(\underline{\xi})$

$$= \int_{t\Omega} e^{-i\underline{\xi}_m\cdot\underline{\eta}}[(2\pi)^{-n} \int_{R_m^t} t^{-n/2}e^{-it^{-1}\underline{z}\cdot\underline{\eta}}u^{\wedge}(\underline{z})d\underline{z}]e^{i\underline{\xi}\cdot\underline{\eta}}d\underline{\eta} .$$

Finally applying $\chi(t,\ell)$ we get $F^{\wedge}(t,\ell,m)u^{\wedge}(\underline{x}) =$

$$\begin{cases} \int_{t\Omega} e^{-i\underline{\xi}_m\cdot\underline{\eta}}(2\pi)^{-n}\int_{R_m^t} t^{-n}u^\wedge(\underline{z})e^{-it^{-1}\underline{z}\cdot\underline{\eta}}d\underline{z}e^{i\underline{\eta}\cdot(t^{-1}\underline{x}+\underline{\xi}_\ell)}d\underline{\eta} & \underline{x}\in R_\ell^t \\ 0 & \underline{x}\notin R_\ell^t \end{cases},$$

$$= \begin{cases} \int_{\Omega} e^{i(\underline{\xi}_\ell-\underline{\xi}_m)\cdot(\underline{t}^{-1}\underline{\eta})}(2\pi)^{-n}\int_{R_m^t} u^\wedge(\underline{z})e^{-i\underline{z}\cdot\underline{\eta}}d\underline{z}e^{i\underline{\eta}\cdot\underline{x}}d\underline{\eta}\,, & \underline{x}\in R_\ell^t\,, \\ 0\,, & \underline{x}\notin R_\ell^t \end{cases}$$

We note that for $F^\wedge(t,\ell,\ell)$ the exponential factor $e^{i(\underline{\xi}_\ell-\underline{\xi}_m)\cdot\underline{\eta}t^{-1}} = 1$ while since the $\underline{\xi}_m$'s are distinct this is not true for $F^\wedge(t,\ell,m)$, $\ell\neq m$. An easy computation shows that if $\underline{u}=[u_1,\ldots,u_p]\in\underset{\sim}{H}^\wedge$ then

$$\underline{F}^\wedge(t)\underline{u}^\wedge = \left[\sum_{m=1}^{p}F^\wedge(t,1,m)u_m^\wedge,\ \ldots,\ \sum_{m=1}^{p}F^\wedge(t,p,m)u_m^\wedge\right].$$

b) The operator $T^\wedge$ on $L^\wedge$ was defined by

$$T^\wedge u^\wedge(\underline{\xi}) = f^\wedge(\underline{\xi})u^\wedge(\underline{\xi})\,.$$

For each $t>0$ we define $\underline{T}^\wedge(t)$ on $\underset{\sim}{H}^\wedge$ by

$$\underline{T}^\wedge(t) = \underline{\chi}(t)T^\wedge\underline{\chi}^*(t)\,.$$

An easy calculation gives the action of $\underline{T}^\wedge(t)$ on an element

$\underline{u}^\wedge = [u_1^\wedge, \ldots, u_p^\wedge]$ of $\underset{\sim}{H}^\wedge$.

$$\underline{T}^\wedge(t)\underline{u}^\wedge = [f_1^t u_1^\wedge, \ldots, f_p^t u_p^\wedge]$$

where
$$f_m^t(\underline{x}) = \begin{cases} f(t^{-1}\underline{x}+\underline{\xi}_m) , & \underline{x} \in R_m^t \\ 0 , & \underline{x} \notin R_m^t , \quad m = 1, \ldots, p . \end{cases}$$

We define $S^\wedge(t,m)$, $m = 1, \ldots, p$ on $\underset{\sim}{H}^\wedge$ by

$$S^\wedge(t,m)u^\wedge = t^\omega L(t^{-1})^{-1} f_m^t u^\wedge$$

$$= g_m^t u^\wedge ,$$

where this equation defines g_m^t . $\underline{S}^\wedge(t)$ is then defined on $\underset{\sim}{H}^\wedge$ by

$$\underline{S}^\wedge(t)\underline{u}^\wedge = [S^\wedge(t,1)u_1^\wedge, \ldots, S^\wedge(t,p)u_p^\wedge] .$$

Note that $\underline{S}^\wedge(t) = t^\omega L(t^{-1})^{-1}\underline{T}^\wedge(t)$ and that for every $t > 0$ $\underline{S}^\wedge(t)$ is a bounded self-adjoint operator on $\underset{\sim}{H}^\wedge$.

c) $\underline{S}^\wedge$ is defined on $\underset{\sim}{H}^\wedge$ as follows. Let

$$g_m(\underline{x}) = |\underline{x}|^\omega \Phi_m(\underline{x}) , \quad m = 1, \ldots, p .$$

Define $S^\wedge(m)$ mapping $\underset{\sim}{H}^\wedge$ to $\underset{\sim}{H}^\wedge$ by

$$S^\wedge(m)u^\wedge = g_m u^\wedge , \quad m = 1, \ldots, p .$$

Finally we set $\underline{S}^\wedge\underline{u}^\wedge = [S^\wedge(1)u_1^\wedge, \ldots, S^\wedge(p)u_p^\wedge]$. $\underline{S}^\wedge$ is a positive self-adjoint operator defined for all $\underline{u}^\wedge \in \underset{\sim}{H}^\wedge$ such that

$$\int_{\underline{R}_n} |g_m(\underline{x})u_m^{\wedge}(\underline{x})|^2 d\underline{x} < \infty \ , \quad m = 1, \ldots, p \ .$$

d) We define a projection $\underline{F}^{\wedge}$ on $\underset{\sim}{H}^{\wedge}$ as follows. Let F be the projection defined in $\underset{\sim}{H}$ by

$$Fu(\underline{y}) = \begin{cases} u(\underline{y}) & \underline{y} \in \Omega \\ 0 & \underline{y} \notin \Omega \end{cases} ,$$

and define $\underline{F}$ on $\underset{\sim}{H}$ by

$$\underline{Fu} = [Fu_1, \ldots, Fu_p] \ .$$

$\underline{F}^{\wedge}$ is defined on $\underset{\sim}{H}^{\wedge}$ by $\underline{F}^{\wedge} = \Psi \underline{F} \Psi^{-1}$. An easy computation gives

$$\underline{F}^{\wedge}\underline{u}^{\wedge}(\underline{x}) = \left[\int_{\Omega} (\psi^{-1}u_1^{\wedge})(\underline{y}) e^{i\underline{x}\cdot\underline{y}} d\underline{y}, \ \ldots, \ \int_{\Omega} (\psi^{-1}u_p^{\wedge})(\underline{y}) e^{i\underline{x}\cdot\underline{y}} d\underline{y} \right] .$$

If

$$\lambda(t,1) \leq \lambda(t,2) \leq \ldots$$

are, for t large, the lowest eigen values of

$$E^{\wedge}(t)TE^{\wedge}(t)\Big|_{E^{\wedge}(t)\underset{\sim}{L}^{\wedge}} ,$$

they are clearly also the lowest eigen values of

$$\underline{F}^{\wedge}(t)\underline{T}^{\wedge}(t)\underline{F}^{\wedge}(t)\Big|_{\underline{F}^{\wedge}(t)\underset{\sim}{H}^{\wedge}} .$$

Similarly

$$t^{\omega}L(t^{-1})^{-1}\lambda(t,1) \leq t^{\omega}L(t^{-1})^{-1}\lambda(t,2) \leq \ldots$$

are the lowest eigen values of the operator

$$\underline{F}^\wedge(t)\underline{S}^\wedge(t)\underline{F}^\wedge(t)\Big|_{\underline{F}^\wedge(t)\underset{\approx}{H}^\wedge} .$$

In sections 3 - 7 of this Chapter we will show that the operators $\underline{F}^\wedge$, $\underline{S}^\wedge$, $\underline{F}^\wedge(t)$, and $\underline{S}^\wedge(t)$ satisfy assumptions i - vi and conditions a. of Chapter I. Thus using the results of that Chapter we find that the eigen values described do in fact exist, and if

$$0 < \mu(1) \leq \mu(2) \leq \dots, \quad \lim_{k \to \infty} \mu(k) = \infty$$

are the eigen values of

$$\underline{S}^\wedge_{\underline{F}^\wedge}$$

repeated according to their multiplicities, then

$$\lim_{t \to \infty} t^{\omega} L(t^{-1})^{-1} \lambda(t,k) = \mu(k) , \quad k = 1, 2, \dots .$$

3. <u>Convergence of</u> $\underline{S}^\wedge(t)^{1/2}$ <u>to</u> $(\underline{S}^\wedge)^{1/2}$

<u>Lemma 3a</u>. If $0 < a_1 \leq y_1 y_2^{-1} \leq a_2 < \infty$ then

$$L(y_1)/L(y_2) \to 1 \quad \text{as} \quad y_1, y_2 \to 0 .$$

Proof. Let $\epsilon > 0$ be given. Then for $y_1 < y_2$, $0 < y_1, y_2 < \alpha(\epsilon)$ we have

(1) $$L(y_1)y_1^{\epsilon} \leq L(y_2)y_2^{\epsilon} ,$$

(2) $$L(y_1)y_1^{-\epsilon} \geq L(y_2)y_2^{-\epsilon} .$$

(1) implies

$$\frac{L(y_1)}{L(y_2)} \leq \frac{y_1^{-\epsilon}}{y_2^{-\epsilon}} .$$

(2) implies $$\frac{L(y_1)}{L(y_2)} \geq \frac{y_1^{\epsilon}}{y_2^{\epsilon}} \ .$$

Hence for $0 < y_1, y_2 < \alpha(\epsilon)$ we have

$$a_1^{\epsilon} \leq L(y_1)/L(y_2) \leq a_2^{\epsilon} \tag{3}$$

and

$$a_1^{-\epsilon} \leq L(y_2)/L(y_1) \leq a_2^{-\epsilon} \ . \tag{4}$$

Since ϵ is arbitrary (3) and (4) imply our result.

Lemma 3b. Given $\epsilon > 0$ and $\eta > 0$ there is a constant $M(\epsilon)$ such that if $0 < y_1, y_2 \leq \eta$ then

$$L(y_1)/L(y_2) \leq M(\epsilon) \left[\left(\frac{y_1}{y_2}\right)^{\epsilon} + \left(\frac{y_1}{y_2}\right)^{-\epsilon}\right] . \tag{5}$$

Proof. Let $\epsilon > 0$ be given. Suppose $0 < y_1 \leq y_2 < \alpha(\epsilon)$. Then

$$L(y_1)y_1^{\epsilon} \leq L(y_2)y_2^{\epsilon} ,$$

which implies that

$$L(y_1)/L(y_2) \leq \left(\frac{y_1}{y_2}\right)^{-\epsilon} .$$

If $0 < y_2 < y_1 \leq \alpha(\epsilon)$ then

$$L(y_2)y_2^{-\epsilon} \geq L(y_1)y_1^{-\epsilon}$$

and so

$$L(y_1)/L(y_2) \leq \left(\frac{y_1}{y_2}\right)^{\epsilon} .$$

Putting these together we have for $0 < y_1, y_2 < \alpha(\epsilon)$

$$L(y_1)/L(y_2) \leq \left(\frac{y_1}{y_2}\right)^{\epsilon} + \left(\frac{y_1}{y_2}\right)^{-\epsilon} .$$

Since $L(y)$ is continuous on $[\alpha(\epsilon),\eta]$ the result clearly follows.

Theorem 3c. For $m = 1, \ldots, p$ $\lim_{t \to \infty} g_m^t(\underline{x}) = g_m(\underline{x})$ for $\underline{x} \neq 0$.

Proof. We recall the definitions of these functions

$$g_m^t(\underline{x}) = \begin{cases} t^{\omega} L(t^{-1})^{-1} f(t^{-1}\underline{x} + \underline{\xi}_m) , & \underline{x} \in R_m^t \\ 0 , & \underline{x} \notin R_m^t \end{cases}$$

and

$$g_m(\underline{x}) = |\underline{x}|^{\omega} \Phi_m(\underline{x}) .$$

Let $\underline{x} \in \underline{R}_n$ be fixed. From the definition of the function $f(\underline{\xi})$ we have

$$g_m^t(\underline{x}) = t^{\omega} L(t^{-1})^{-1}\left(\Phi_m(\underline{x}/|\underline{x}|) L(t^{-1}|\underline{x}|) |t|^{-\omega} |\underline{x}|^{\omega} + o(|t|^{-\omega} |\underline{x}|^{\omega} L(t^{-1}|\underline{x}|)\right)$$

$$= L(t^{-1})^{-1} |\underline{x}|^{\omega} L(t^{-1}|\underline{x}|) \Phi_m(\underline{x}/|\underline{x}|) + o(|\underline{x}|^{\omega} L(t^{-1})^{-1} L(|\underline{x}| t^{-1}) .$$

By Lemma 3a

$$L(t^{-1}|\underline{x}|) L(t^{-1})^{-1} \to 1 \quad \text{as} \quad t \to \infty .$$

Therefore

$$\lim_{t \to \infty} g_m^t(\underline{x}) = g_m(\underline{x}) .$$

Theorem 3d. i. Given $\epsilon > 0$ there exists $A(\epsilon) > 0$ such that for all $\underline{x} \in \underline{R}_n$ and all $t \geq 1$

$$g_m^t(\underline{x}) \leq A(\epsilon)|t|^{\omega+\epsilon}, \; m = 1, \ldots, p \; ;$$

ii. given $\epsilon > 0$ there is a constant $A(\epsilon)$, independent of $\underline{x} \in \underline{R}_n$ such that

$$0 \leq g_m^t(\underline{x}) \leq A(\epsilon)[|\underline{x}|^{\epsilon} + |\underline{x}|^{-\epsilon}]|\underline{x}|^{\omega} ,$$

$m = 1, \ldots, p$. Note: $g_m^t(\underline{x}) = 0$ for $\underline{x} \notin R_m^t$.

Proof. We have $g_m^t(\underline{x}) = t^{\omega}L(t^{-1})^{-1}f(t^{-1}\underline{x}+\underline{\xi}_m)$, $\underline{x} \in R_m^t$. Since f is bounded we clearly have i. By assumption

$$f(\underline{\xi}) = \Phi_m(\underline{\xi}-\underline{\xi}_m)L(|\underline{\xi}-\underline{\xi}_m|)|\underline{\xi}-\underline{\xi}_m|^{\omega} + o(L|\underline{\xi}-\underline{\xi}_m|)|\underline{\xi}-\underline{\xi}_m|^{\omega})$$

as $\underline{\xi} \to \underline{\xi}_m$. It follows from this that we can find a constant M and a neighborhood, N_m, of $\underline{\xi}_m$, $N_m = \{\underline{\xi}: |\underline{\xi}-\underline{\xi}_m| < \eta\}$ such that

$$0 \leq f(\underline{\xi}) \leq ML(|\underline{\xi}-\underline{\xi}_m|)|\underline{\xi}-\underline{\xi}_m|^{\omega}$$

for $\underline{\xi} \in N_m$. Thus if $\underline{x} \in \sigma(t,m)N_m \subset R_m^t$ we have

$$0 \leq f(t^{-1}\underline{x}+\underline{\xi}_m) \leq ML(t^{-1}|\underline{x}|)t^{-\omega}|\underline{x}|^{\omega} .$$

Therefore

$$0 \leq g_m^t(\underline{x}) \leq ML(t^{-1})^{-1}L(t^{-1}|\underline{x}|)|\underline{x}|^{\omega} .$$

Applying Lemma 3b we get

$$0 \le g_m^t(\underline{x}) \le M \cdot M(\epsilon)[|\underline{x}|^{\epsilon} + |\underline{x}|^{-\epsilon}]|\underline{x}|^{\omega}$$

for $\underline{x} \in \sigma(t,m)N_m$. From i. we have

$$0 \le g_m^t(\underline{x}) \le A|\underline{x}|^{\omega+\epsilon}$$

for all $\underline{x} \in \underline{R}_n \backslash \sigma(t,m)N_m$, and ii follows.

We now recall the definitions of the following operators on $\underset{\sim}{H^{\wedge}}$. For $m = 1, \ldots, p$

$$S^{\wedge}(t,m)u^{\wedge}(\underline{x}) = g_m^t(\underline{x})u^{\wedge}(\underline{x})$$

and

$$S^{\wedge}(m)u^{\wedge}(\underline{x}) = g_m(\underline{x})u^{\wedge}(\underline{x}) \ .$$

For δ , $0 < \delta < 1$ let

$$a(\delta,\underline{x}) = \begin{cases} 1 & \delta < |\underline{x}| < \delta^{-1} \\ 0 & \text{otherwise} \end{cases}$$

Corollary 3e. For $u^{\wedge} \in \underset{\sim}{D}[S^{\wedge}(m)^{1/2}]$ and $0 < \delta < 1$ we have

$$\lim_{t \to \infty} \| [S^{\wedge}(t,m)^{1/2} - S^{\wedge}(m)^{1/2}]a(\delta,\underline{x})u^{\wedge}(\underline{x})\|_{\underset{\sim}{H^{\wedge}}} = 0$$

Proof. This is a direct application of 3c, 3d and the Lebesgue dominated convergence theorem.

Theorem 3f. $(\underline{S}^{\wedge})^{1/2}$ is the closure of the strong limit of $(\underline{S}^{\wedge}(t))^{1/2}$ as $t \to \infty$.

Proof. Let $\underline{u}^{\wedge} = [u_1^{\wedge}, \ldots, u_p^{\wedge}] \in \underset{\sim}{D}((\underline{S}^{\wedge})^{1/2})$ and $\epsilon > 0$ be given.

Consider

$$\underline{u}^\wedge_\delta(\underline{x}) = [a(\delta,\underline{x})u^\wedge_1(\underline{x}), \ldots, a(\delta,\underline{x})u^\wedge_p(\underline{x})] .$$

For $\delta > 0$ sufficiently small

$$\|\underline{u}^\wedge - \underline{u}^\wedge_\delta\|^2_{\underline{\underline{H}}^\wedge} = \sum_{m=1}^{p} \int_{\underline{R}_n} |u^\wedge_m(\underline{x})|^2 |1-a(\delta,\underline{x})|^2 d\underline{x} < \varepsilon$$

and

$$\|(\underline{S}^\wedge)^{1/2}(\underline{u}^\wedge - \underline{u}^\wedge_\delta)\|_{\underline{\underline{H}}^\wedge} = \sum_{m=1}^{p} \int_{\underline{R}_n} g^t_m(\underline{x}) |u^\wedge_m(\underline{x})|^2 |1-a(\delta,\underline{x})|^2 d\underline{x} < \varepsilon .$$

It follows directly from Corollary 3e that

$$\underline{S}^\wedge(t)^{1/2}\underline{u}^\wedge_\delta \to (\underline{S}^\wedge)^{1/2}\underline{u}^\wedge_\delta \quad \text{in } \underline{\underline{H}}^\wedge$$

and this completes the proof.

4. Convergence of $\underline{F}^\wedge(t)$ to $\underline{F}^\wedge$

Throughout the present section we suppose that Ω is any set of finite positive measure for which $\partial\Omega$ has measure 0. While for the purposes of this chapter we need only that Ω has finite positive measure and is locally star-shaped, the present more general results are required in Chapter VII.

Let $\underset{\sim}{D}(\Omega)$ be the set of functions in $\underset{\sim}{H}$ which are infinitely differentiable with compact support contained in Ω. Then set $\underset{\sim}{D}^\wedge(\Omega) = \phi\underset{\sim}{D}(\Omega)$. $\underset{\approx}{D}(\Omega)(\underset{\approx}{D}^\wedge(\Omega))$ is defined as the collection of elements of $\underset{\approx}{H}(\underset{\approx}{H}^\wedge)$ with each component an element of $\underset{\sim}{D}(\Omega)(\underset{\sim}{D}^\wedge(\Omega))$. We also define corresponding sets for Ω', the complement of Ω, and denote them by $\underset{\sim}{D}(\Omega')$, $\underset{\sim}{D}^\wedge(\Omega')$, etc. Since the boundary of a locally star-shaped set has measure zero the vector space

$\underset{\sim}{D}(\Omega) + \underset{\sim}{D}(\Omega')(\underset{\sim}{D}^{\wedge}(\Omega)+\underset{\sim}{D}^{\wedge}(\Omega'))$ is dense in $\underset{\sim}{H}(\underset{\sim}{H}^{\wedge})$.

Lemma 4a. If $u^{\wedge}(\underline{x}) \in \underset{\sim}{H}^{\wedge}$ and $u^{\wedge}(\underline{x}) = O(|\underline{x}|^{-s})$ for every positive number s then for ℓ , $m = 1, \ldots, p$, $\ell \neq m$ we have

$$\|u^{\wedge}(\underline{x}+t(\underline{\xi}_{\ell}-\underline{\xi}_m))|_{R_{\ell}^t}\|_{\infty} = O(t^{-r}) \quad \text{as} \quad t \to \infty ,$$

and

$$\|u^{\wedge}(\underline{x}+t(\underline{\xi}_{\ell}-\underline{\xi}_m))|_{R_{\ell}^t}\|_{\underset{\sim}{H}^{\wedge}} = O(t^{-r}) \quad \text{as} \quad t \to \infty ,$$

for every positive number r .

Proof. We first claim that the sets $R_{\ell}^t + t(\underline{\xi}_{\ell}-\underline{\xi}_m)$ "recede to infinity" as $t \to \infty$. More specifically, there is a $q > 0$ such that

$$\text{dist}(R_{\ell}^t+t(\underline{\xi}_{\ell}-\underline{\xi}_m),\underline{0}) \geq tq .$$

We show that $\underline{0} \notin R_{\ell}^1 + (\underline{\xi}_{\ell}-\underline{\xi}_m)$. If $R_{\ell} = [\underline{a}_{\ell},\underline{b}_{\ell}]$ where $\underline{a}_{\ell} = (a_{\ell 1}, \ldots, a_{\ell n})$, $\underline{b}_{\ell} = (b_{\ell 1}, \ldots, b_{\ell n})$ then

$$R_{\ell}^1 = [\underline{a}_{\ell}-\underline{\xi}_{\ell},\underline{b}_{\ell}-\underline{\xi}_{\ell}] .$$

If $\underline{0} \in R_{\ell}^1 + (\underline{\xi}_{\ell}-\underline{\xi}_m)$ then $\underline{\xi}_m - \underline{\xi}_{\ell} \in R_{\ell}^1$; that is, for $1 \leq j \leq n$

$$a_{\ell j} - \xi_{\ell j} \leq \xi_{mj} - \xi_{\ell j} \leq b_{\ell j} - \xi_{\ell j}$$

or equivalently

$$a_{\ell j} \leq \xi_{mj} \leq b_{\ell j} .$$

This implies that $\underline{\xi}_m \in R_{\ell}$ which is a contradiction. It is now clear that our claim is true with

$$q = \mathrm{dist}(R^1_\ell + \underline{\xi}_\ell - \underline{\xi}_m, \underline{0}) \ .$$

Our first assertion is an immediate consequence. Since

$S^t_\ell = \{\underline{z}: |\underline{z}| < qt\}$ is disjoint from $R^t_\ell + t(\underline{\xi}_\ell - \underline{\xi}_m)$, we have

$$\int_{R^t_\ell} |u^\wedge(\underline{x} + t(\underline{\xi}_\ell - \underline{\xi}_m))|^2 d\underline{x} = \int_{R^t_\ell + t(\underline{\xi}_\ell - \underline{\xi}_m)} |u^\wedge(\underline{y})|^2 d\underline{y} \ ,$$

$$\leq \int_{|\underline{y}| \geq qt} |u^\wedge(\underline{y})|^2 d\underline{y} \ .$$

In view of the fact that

$$u^\wedge(\underline{x}) = O(|\underline{x}|^{-s}) \quad \text{as} \quad |\underline{x}| \to \infty$$

for every positive integer s, our second assertion follows.

Theorem 4b. Let s be any positive number. We have:

i. if $u^\wedge \in \underset{\sim}{D}^\wedge(\Omega)$ then for $\ell = 1, \ldots, p$

$$\|F^\wedge(t,\ell,\ell)u^\wedge - u^\wedge\|_{\underset{\sim}{H}^\wedge} = o(t^{-s}) \quad \text{as} \quad t \to +\infty \ ,$$

$$\|F^\wedge(t,\ell,\ell)u^\wedge - u^\wedge\|_\infty = o(t^{-s}) \quad \text{as} \quad t \to +\infty \ ;$$

ii. if $u^\wedge \in D^\wedge(\Omega')$ then for $\ell = 1, \ldots, p$

$$\|F^\wedge(t,\ell,\ell)u^\wedge\|_{\underset{\sim}{H}^\wedge} = o(t^{-s}) \quad \text{as} \quad t \to +\infty \ ,$$

$$\|F^\wedge(t,\ell,\ell)u^\wedge\|_\infty = o(t^{-s}) \quad \text{as} \quad t \to +\infty \ ;$$

iii. if $u^\wedge \in D^\wedge(\Omega)$ then for $1 \leq \ell, m \leq p, \ell \neq m$

$$\|F^\wedge(t,\ell,m)u^\wedge\|_{\underset{\sim}{H}^\wedge} = o(t^{-s}) \quad \text{as} \quad t \to +\infty \ ,$$

$$\|F^{\wedge}(t,\ell,m)u^{\wedge}\|_{\infty} = o(t^{-s}) \quad \text{as} \quad t \to +\infty ;$$

iv. if $u^{\wedge} \in \underset{\sim}{D}^{\wedge}(\Omega')$ then for $1 \leq \ell,\ m \leq p,\ \ell \neq m$

$$\|F^{\wedge}(t,\ell,m)u^{\wedge}\|_{\underset{\sim}{H}^{\wedge}} = o(t^{-s}) \quad \text{as} \quad t \to +\infty ,$$

$$\|F^{\wedge}(t,\ell,m)u^{\wedge}\|_{\infty} = o(t^{-s}) \quad \text{as} \quad t \to +\infty .$$

Proof. i. Let $u^{\wedge} \in \underset{\sim}{D}^{\wedge}(\Omega)$. We have for $\underline{x} \in R_{\ell}^{t}$

$$F^{\wedge}(t,\ell,\ell)u^{\wedge}(\underline{x}) = \int_{\Omega} e^{i\underline{x}\cdot\underline{\eta}}d\underline{\eta}(2\pi)^{-n}\int_{R_{\ell}^{t}} u^{\wedge}(\underline{z})e^{-i\underline{z}\cdot\underline{\eta}}d\underline{\eta} ,$$

$$= \int_{\Omega} e^{i\underline{x}\cdot\underline{\eta}}u(\underline{\eta})d\underline{\eta} - \int_{\Omega} e^{i\underline{x}\cdot\underline{\eta}}d\underline{\eta}(2\pi)^{-n}\int_{\underline{R}_n\setminus R_{\ell}^{t}} u^{\wedge}(\underline{z})e^{-i\underline{z}\cdot\underline{\eta}}d\underline{z} ,$$

$$= u^{\wedge}(\underline{x}) - t^{-s}u_t^{\wedge}(\underline{x})$$

where

(1) $$u_t^{\wedge}(\underline{x}) = \int_{\Omega} e^{i\underline{x}\cdot\underline{\eta}}d\underline{\eta}(2\pi)^{-n}\int_{\underline{R}_n} \{t^s u^{\wedge}(\underline{z})\Big|_{\underline{R}_n\setminus R_{\ell}^{t}}\}e^{-i\underline{z}\cdot\underline{\eta}}d\underline{z} .$$

Since $u^{\wedge}(\underline{z}) = O(|z|^{-r})$ as $|\underline{z}| \to \infty$ for every $r > 0$ it follows that

(2) $$\|t^s u^{\wedge}(\underline{z})\Big|_{\underline{R}_n\setminus R_{\ell}^{t}}\|_{\underset{\sim}{H}^{\wedge}} = o(1) \text{ and } \|t^s u^{\wedge}(\underline{z})\Big|_{\underline{R}_n\setminus R^{t}}\|_{1} = o(1) \quad \text{as} \quad t \to \infty .$$

Two applications of Plancherel's theorem show that

$$\|u_t^{\wedge}\|_{\underset{\sim}{H}^{\wedge}} = o(1) \quad \text{as} \quad t \to \infty ,$$

while an even more elementary argument using the fact that Ω has finite measure shows that

$$\|u_t^\wedge\|_\infty = o(1) \quad \text{as} \quad t \to \infty .$$

Since

$$F^\wedge(t,\ell,\ell)u^\wedge(\underline{x}) - u^\wedge(\underline{x}) = \{t^{-s}u_t^\wedge(\underline{x})\Big|_{R_\ell^t}\} - \{u^\wedge(\underline{x})\Big|_{\underline{R}_n \setminus R_\ell^t}\}$$

our assertions follow.

ii. Let $u^\wedge \in \underset{\sim}{D}^\wedge(\Omega')$. Since $\mathrm{supp}(\psi^{-1}u) \subset \Omega'$ we have for $\underline{x} \in R_\ell^t$

$$F^\wedge(t,\ell,\ell)u^\wedge(\underline{x}) = -t^{-s}u_t^\wedge(\underline{x})$$

where $u_t^\wedge(\underline{x})$ is defined as in (1). Using (2) and

$$F^\wedge(t,\ell,\ell)u^\wedge(\underline{x}) = -t^{-s}u_t^\wedge(\underline{x})\Big|_{R_\ell^t}$$

we can complete our arguments as before.

iii. Let $u^\wedge \in \underset{\sim}{D}^\wedge(\Omega)$. We have for $\underline{x} \in R_\ell^t$

$$F^\wedge(t,\ell,m)u^\wedge(\underline{x}) = \int_\Omega e^{i\underline{\eta}\cdot t(\underline{\xi}_\ell - \underline{\xi}_m)} e^{i\underline{x}\cdot\underline{\eta}} d\underline{\eta}(2\pi)^{-n} \int_{R_m^t} u^\wedge(\underline{z})e^{-i\underline{z}\cdot\underline{\eta}}d\underline{z} ,$$

$$= \int_\Omega e^{i\underline{\eta}\cdot(\underline{x} + t(\underline{\xi}_\ell - \underline{\xi}_m))}(\psi^{-1}u^\wedge)(\underline{\eta})d\underline{\eta}$$

$$- \int_\Omega e^{i\underline{\eta}\cdot(\underline{x} + t(\underline{\xi}_\ell - \underline{\xi}_m))} d\underline{\eta}(2\pi)^{-n} \int_{\underline{R}_n \setminus R_m^t} u^\wedge(\underline{z})e^{-i\underline{z}\cdot\underline{\eta}}d\underline{z} ,$$

$$= u^\wedge(\underline{x} + t(\underline{\xi}_\ell - \underline{\xi}_m)) - t^{-s}u_t^\wedge(\underline{x} + t(\underline{\xi}_\ell - \underline{\xi}_m))$$

where

$$u_t^{\wedge}(\underline{x}) = \int_{\Omega} e^{i\underline{\eta}\cdot\underline{x}} d\underline{\eta} (2\pi)^{-n} \int_{\underline{R}_n} \{t^s u^{\wedge}(\underline{z})\Big|_{\underline{R}_n \backslash R_m^t}\} e^{-i\underline{z}\cdot\underline{\eta}} d\underline{z} .$$

Using (2), with ℓ replaced by m, we see that

$$\|u_t^{\wedge}(\underline{x}+t(\underline{\xi}_{\ell}-\underline{\xi}_m))\|_{\underset{\sim}{H}^{\wedge}} = o(1) , \|u_t^{\wedge}(\underline{x}+t(\underline{\xi}_{\ell}-\underline{\xi}_m))\|_{\infty} = o(1) \quad \text{as} \quad t \to \infty .$$

That

$$\|u^{\wedge}(\underline{x}+t(\underline{\xi}_{\ell}-\underline{\xi}_m))\Big|_{R_{\ell}^t}\|_{\underset{\sim}{H}^{\wedge}} = o(t^{-r}) \quad \text{as} \quad t \to \infty ,$$

and

$$\|u^{\wedge}(\underline{x}+t(\underline{\xi}_{\ell}-\underline{\xi}_m))\Big|_{R_{\ell}^t}\|_{\infty} = o(t^{-r}) \quad \text{as} \quad t \to \infty ,$$

is the assertion of Lemma 4a.

iv. The proof of iv is parallel to that if iii.

Corollary 4c. $\underline{F}^{\wedge}$ is the strong limit of $\underline{F}^{\wedge}(t)$ as $t \to \infty$.

Proof. If $\underline{u}^{\wedge} \in \underset{\sim}{D}^{\wedge}(\Omega) + \underset{\sim}{D}^{\wedge}(\Omega')$ then it is an immediate consequence of Lemma 4b that

$$\|\underline{F}^{\wedge}(t)\underline{u}^{\wedge} - \underline{F}^{\wedge}\underline{u}^{\wedge}\|_{\underset{\sim}{H}^{\wedge}} \to 0 \quad \text{as} \quad t \to \infty .$$

Since, as noted at the beginning of this section, this is a dense subset of $\underset{\sim}{H}^{\wedge}$ our conclusion follows.

5. Convergence of $\underline{S}^{\wedge}(t)^{1/2}\underline{F}^{\wedge}(t)$ to $(\underline{S}^{\wedge})^{1/2}$ on $\underset{\sim}{M}^{\wedge}$, Part I

Henceforth Ω is as in i. of §1.

Lemma 5a. If $u^{\wedge} \in \underset{\sim}{D}^{\wedge}(\Omega)$ then for $1 \leq \ell \leq p$

$$\|S^{\wedge}(t,\ell)^{1/2} F^{\wedge}(t,\ell,\ell)u^{\wedge} - S^{\wedge}(\ell)^{1/2} F^{\wedge} u^{\wedge}\|_{\underset{\sim}{H}^{\wedge}} \to 0 \quad \text{as} \quad t \to \infty .$$

Proof. We first observe that since $u^\wedge \in \underset{\sim}{D}^\wedge(\Omega)$ we have $F^\wedge u^\wedge = u^\wedge$. Thus by the triangle inequality

$$(1) \qquad \|S^\wedge(t,\ell)^{1/2}F^\wedge(t,\ell,\ell)u^\wedge - S^\wedge(\ell)^{1/2}F^\wedge u^\wedge\|_{\underset{\sim}{H}^\wedge}$$

$$\leq \|S^\wedge(t,\ell)^{1/2}\{F^\wedge(t,\ell,\ell)u^\wedge - u^\wedge\}\|_{\underset{\sim}{H}^\wedge} + \|\{S^\wedge(t,\ell)^{1/2} - S^\wedge(\ell)^{1/2}\}u^\wedge\|_{\underset{\sim}{H}^\wedge} .$$

From conclusion i of Theorem 3d we have

$$\sup_{\underline{x}} g_\ell^t(\underline{x}) = 0(t^{\omega+\epsilon}) .$$

Therefore from i of Theorem 4b we have

$$(2) \qquad \|S^\wedge(t,\ell)^{1/2}\{F^\wedge(t,\ell,\ell)u^\wedge - u^\wedge\}\|_{\underset{\sim}{H}^\wedge}$$

$$= \{(2\pi)^{-n}\int_{\underline{R}_n} g_\ell^t(\underline{x})|F^\wedge(t,\ell,\ell)u^\wedge(\underline{x}) - u^\wedge(\underline{x})|^2 d\underline{x}\}^{1/2}$$

$$= 0(t^{\omega+\epsilon})\|F^\wedge(t,\ell,\ell)u^\wedge - u^\wedge\|_{\underset{\sim}{H}^\wedge}$$

$$= 0(t^{-s}) \quad \text{as} \quad t \to \infty \quad \text{for any positive integer} \quad s .$$

On the other hand we have

$$\|\{S^\wedge(t,\ell)^{1/2} - S^\wedge(\ell)^{1/2}\}u^\wedge\|_{\underset{\sim}{H}^\wedge}$$

$$= \{(2\pi)^{-n}\int_{\underline{R}_n} |g_\ell^t(\underline{x})^{1/2} - g(\underline{x})^{1/2}|^2|u^\wedge(\underline{x})|^2 d\underline{x}\}^{1/2} .$$

Furthermore

$$|g_\ell^t(\underline{x})^{1/2} - g_\ell(\underline{x})^{1/2}|^2 \leq 2g_\ell^t(\underline{x}) + 2g_\ell(\underline{x})$$

and by Theorem 3d given $\epsilon > 0$ there exists a constant $A(\epsilon)$ such that

$$g_{\ell}^{t}(\underline{x}) \leq A(\epsilon)[|\underline{x}|^{\epsilon} + |\underline{x}|^{-\epsilon}]|\underline{x}|^{\omega}$$

for all $\underline{x} \in \underline{R}_n$. By definition

$$g_{\ell}(\underline{x}) = \Phi_{\ell}(\underline{x})|\underline{x}|^{\omega}$$

and since $\Phi_{\ell}(\underline{x}) \leq K$ for some $K > 0$

$$g_{\ell}(\underline{x}) \leq K|\underline{x}|^{\omega} .$$

Moreover from Theorem 3c we have for all $\underline{x} \in \underline{R}_n$

$$\lim_{t \to \infty} g_{\ell}^{t}(\underline{x}) = g_{\ell}(\underline{x}) .$$

Since $u\hat{}(\underline{x}) \in \underset{\sim}{D}\hat{}(\Omega)$

$$u\hat{}(\underline{x}) = 0(|\underline{x}|^{-s}) \quad \text{as} \quad |\underline{x}| \to \infty$$

for every positive number s . We may therefore apply the Lebesgue dominated convergence theorem to deduce that

$$\text{(3)} \qquad \|\{S\hat{}(t,\ell)^{1/2} - S\hat{}(\ell)^{1/2}\}u\hat{}\|_{\underset{\sim}{H}\hat{}} \to 0$$

as $t \to \infty$. (1), (2) and (3) establish the lemma.

Lemma 5b. If $u\hat{} \in \underset{\sim}{D}\hat{}(\Omega)$ then for $1 \leq \ell$, $m \leq p$, $\ell \neq m$

$$\|S\hat{}(t,\ell)^{1/2}F\hat{}(t,\ell,m)u\hat{}\|_{\underset{\sim}{H}\hat{}} \to 0 \quad \text{as} \quad t \to \infty .$$

Proof. We have

$$\|S\hat{}(t,\ell)^{1/2}F\hat{}(t,\ell,m)u\hat{}\|_{\underset{\sim}{H}\hat{}} = \{(2\pi)^{-n}\int_{\underline{R}_n} g_{\ell}(\underline{x})^{1/2}|F\hat{}(t,\ell,m)u\hat{}(\underline{x})|^2 d\underline{x}\}^{1/2} .$$

From conclusion i of Theorem 3d we have

$$\sup_{\underline{x}} g_{\ell}^{t}(\underline{x}) = O(t^{\omega+\varepsilon}) .$$

Thus from iii of Theorem 4b we have for $t > \eta^{-1}$

$$\|S^{\wedge}(t,\ell)^{1/2}F^{\wedge}(t,\ell,m)u^{\wedge}\|_{\underset{\sim}{H}^{\wedge}} = O(t^{\omega+\varepsilon})\|F^{\wedge}(t,\ell,m)u^{\wedge}\|_{\underset{\sim}{H}^{\wedge}} = O(t^{-s})$$

as $t \to \infty$ for any positive integer s.

Theorem 5c. If $\underline{u} \in \underset{\sim}{D}^{\wedge}(\Omega)$ then

(4) $$\|\underline{S}^{\wedge}(t)^{1/2}\underline{F}^{\wedge}(t)\underline{u}^{\wedge} - (\underline{S}^{\wedge})^{1/2}\underline{F}^{\wedge}\underline{u}^{\wedge}\|_{\underset{\sim}{H}^{\wedge}} \to 0 \quad \text{as} \quad t \to \infty .$$

Proof. We have

$$\|\underline{S}^{\wedge}(t)^{1/2}\underline{F}(t)\underline{u}^{\wedge} - (\underline{S}^{\wedge})^{1/2}\underline{F}^{\wedge}\underline{u}^{\wedge}\|_{\underset{\sim}{H}^{\wedge}}^{2}$$

$$= \sum_{\ell=1}^{p} \|S^{\wedge}(t,\ell)^{1/2}\{\sum_{m=1}^{p} F^{\wedge}(t,\ell,m)u_m^{\wedge}\} - S^{\wedge}(\ell)^{1/2}u_{\ell}^{\wedge}\|_{\underset{\sim}{H}^{\wedge}}^{2}$$

$$\leq \sum_{\ell=1}^{p} \|S^{\wedge}(t,\ell)^{1/2}F^{\wedge}(t,\ell,\ell)u_{\ell}^{\wedge} - S^{\wedge}(\ell)^{1/2}F^{\wedge}u_{\ell}^{\wedge}\|_{\underset{\sim}{H}^{\wedge}}^{2}$$

$$+ \sum_{\ell \neq m} \|S^{\wedge}(t,\ell)^{1/2}F^{\wedge}(t,\ell,m)u_m^{\wedge}\|_{\underset{\sim}{H}^{\wedge}}^{2} .$$

These last two sums tend to 0 as $t \to \infty$ by Lemma 5a and Lemma 5b. This completes the proof.

6. Convergence of $\underline{S}^{\wedge}(t)^{1/2}\underline{F}^{\wedge}(t)$ to $(\underline{S}^{\wedge})^{1/2}$ on $\underset{\sim}{M}$, Part II

For $1 \leq m \leq p$ we set

$$\underset{\sim}{S}^{\wedge}(m) = \{u^{\wedge} \in F^{\wedge}H^{\wedge} : u^{\wedge} \in \underset{\sim}{D}[S^{\wedge}(m)^{1/2}]^{\wedge}\} \ ,$$

$$\underset{\sim}{M}^{\wedge}(m) = \{\text{closure of } \ \underset{\sim}{S}^{\wedge}(m) \ \text{ in } \ F^{\wedge}\underset{\sim}{H}^{\wedge}\} \ ,$$

$$\underset{\sim}{S}^{\wedge} = \{[u_1^{\wedge}, \ldots, u_p^{\wedge}] : u_m^{\wedge} \in \underset{\sim}{S}_m^{\wedge}, \ 1 \le m \le p\} \ ,$$

$$\underset{\sim}{M}^{\wedge} = \{[u_1^{\wedge}, \ldots, u_p^{\wedge}] : u_m^{\wedge} \in \underset{\sim}{M}_m^{\wedge}, \ 1 \le m \le p\} \ .$$

Theorem 6a. Given $\underline{u}^{\wedge} \in \underset{\sim}{S}^{\wedge}$ and $\epsilon > 0$ there exists $\underline{w}^{\wedge} \in \underset{\sim}{D}^{\wedge}(\Omega)$ such that

$$\|\underline{u}^{\wedge} - \underline{w}^{\wedge}\|_{\underset{\sim}{H}^{\wedge}} < \epsilon \quad \text{and} \quad \|(S^{\wedge})^{1/2}(\underline{u}^{\wedge} - \underline{w}^{\wedge})\|_{\underset{\sim}{H}^{\wedge}} < \epsilon \ .$$

The demonstration splits into two rather distinct parts.

Part 1. We show here that given $\underline{u}^{\wedge} \in \underset{\sim}{S}^{\wedge}$ and $\epsilon > 0$ there exists a finite set of functions $\underline{u}_r^{\wedge}$ $r = 1, \ldots, R$ in $\underset{\sim}{S}^{\wedge}$ such that

$$\|\underline{u}^{\wedge} - \sum_1^R \underline{u}_r^{\wedge}\|_{\underset{\sim}{H}^{\wedge}} < \epsilon, \ \|(\underline{S}^{\wedge})^{1/2}(\underline{u}^{\wedge} - \sum_1^R \underline{u}_r^{\wedge})\|_{\underset{\sim}{H}^{\wedge}} < \epsilon$$

and such that each $\underline{u}_r^{\wedge}$ vanishes outside a bounded star-shaped subset Ω_r of Ω .

Part 2. We show here that given $\underline{u}^{\wedge} \in \underset{\sim}{S}^{\wedge}$ such that $\underline{u}^{\wedge}$ vanishes outside a bounded star-shaped subset of Ω , and given $\epsilon > 0$ there exists $\underline{w}^{\wedge} \in \underset{\sim}{D}^{\wedge}(\Omega)$ such that

$$\|\underline{u}^{\wedge} - \underline{w}^{\wedge}\|_{\underset{\sim}{H}^{\wedge}} < \epsilon, \ \|(\underline{S}^{\wedge})^{1/2}(\underline{u}^{\wedge} - \underline{w}^{\wedge})\|_{\underset{\sim}{H}^{\wedge}} < \epsilon \ .$$

Proof of Part 1. Let $\underset{\sim}{K}^{\wedge}$ be the set of all functions u on $\underline{R}_n$ for which

$$(2\pi)^{-n} \int_{\underline{R}_n} (1+|\underline{x}|)^{\omega} |u(\underline{x})|^2 dx < \infty .$$

If for $u , v \in \underset{\sim}{K}^{\wedge}$ we set

$$(u|v) = (2\pi)^{-n} \int_{\underline{R}_n} (1+|\underline{x}|^{\omega} u(\underline{x})\overline{v(\underline{x})} d\underline{x} ,$$

$$|||u||| = (u|u)^{1/2} ,$$

then $\underset{\sim}{K}^{\wedge}$ is a Hilbert space. The point of introducing $\underset{\sim}{K}^{\wedge}$ lies in the fact that $u^{\wedge} \in \underset{\sim}{S}^{\wedge}_m$ if and only if $\underset{\sim}{u}^{\wedge} \in \underset{\sim}{K}^{\wedge}$ and

$$A\,|||u^{\wedge}||| \leq \|(S^{\wedge}_m)^{1/2} u^{\wedge}\|_{H^{\wedge}} \leq A\,|||u^{\wedge}||| \qquad m = 1, \ldots, p$$

since each $\Phi_m(\underline{x})$ satisfies inequalities of the form

$$0 < A \leq \Phi_m(\underline{x}) \leq a < \infty \qquad \underline{x} \in \underline{R}_n$$

(Throughout we use A for <u>various</u> positive constants).

a. Let $\underline{u}^{\wedge} \in \underset{\sim}{S}^{\wedge}$. We assert that given $\epsilon > 0$ there is a function $\underline{v}^{\wedge} \in \underset{\approx}{S}^{\wedge}$ such that

$$\|\underline{u}^{\wedge}-\underline{v}^{\wedge}\| < \epsilon , \quad |||\underline{u}^{\wedge}-\underline{v}^{\wedge}||| < \epsilon$$

and such that

$$v_k(\underline{y}) = 0 \qquad \underline{y} \notin \Omega_b \qquad k = 1, \ldots, p$$

where Ω_b is a bounded subset of Ω .

Let $\alpha(\underline{y})$ be a C^{∞} function on $\underline{R}_n$ satisfying

(1) $$\operatorname{supp}(\alpha) \subset \{\underline{y} : |\underline{y}| \leq 1\}$$

(2) $$\alpha(\underline{0}) = 1$$

(3) $$\alpha^\wedge(\underline{x}) \geq 0 \; .$$

Let $\alpha_\eta(\underline{y}) = \alpha(\eta\underline{y})$, $\eta > 0$. We have

$$|||\alpha_\eta^\wedge * u_m^\wedge|||^2 = (2\pi)^{-n} \int_{\underline{R}_n} (1+|\underline{x}|)^\omega |\alpha_\eta^\wedge * u_m^\wedge(\underline{x})|^2 d\underline{x} \; ,$$

$$= (2\pi)^{-n} \int_{\underline{R}_n} (1+|\underline{x}|)^\omega d\underline{x} \left| (2\pi)^{-n}\eta^{-n} \int_{\underline{R}_n} \alpha^\wedge(\eta^{-1}(\underline{x}-\underline{z}))u_m^\wedge(\underline{z})d\underline{z}\right|^2 ,$$

$$\leq (2\pi)^{-n} \int_{\underline{R}_n} (1+|\underline{x}|)^\omega d\underline{x} (2\pi)^{-n}\eta^{-n} \int_{\underline{R}_n} |\alpha^\wedge(\eta^{-1}(\underline{x}-\underline{z}))| \, |u_m^\wedge(\underline{z})|^2 d\underline{z}$$

$$\times (2\pi)^{-n}\eta^{-n} \int_{\underline{R}_n} |\alpha^\wedge(\eta^{-1}(\underline{x}-\underline{z}))| d\underline{z} \; .$$

Now,

$$(2\pi)^{-n} \int_{\underline{R}_n} \eta^{-n} |\alpha^\wedge(\eta^{-1}(\underline{x}-\underline{z}))| d\underline{z} = (2\pi)^{-n} \int_{\underline{R}_n} \alpha^\wedge(\underline{y})d\underline{y} = \alpha(0) = 1 \; ,$$

where we have made the change of variables $\eta^{-1}(\underline{x}-\underline{z}) = \underline{y}$ and used (2) and (3). Therefore since $(1+|\underline{x}|)^\omega$ is submultiplicative

$$|||\alpha_\eta^\wedge * u^\wedge|||^2$$

$$\leq (2\pi)^{-n} \int_{\underline{R}_n} (1+|\underline{x}|)^\omega d\underline{x} (2\pi)^{-n}\eta^{-n} \int_{\underline{R}_n} \alpha^\wedge(\eta^{-1}(\underline{x}-\underline{z}))|u_m^\wedge(\underline{z})|^2 d\underline{z} \; ,$$

$$\leq (2\pi)^{-n} \int_{\underline{R}_n} (1+|\underline{z}|)^\omega |u_m^\wedge(\underline{z})|^2 dz (2\pi)^{-n}\eta^{-n} \int_{\underline{R}_n} (1+|\underline{x}-\underline{z}|)^\omega \alpha^\wedge(\eta^{-1}(\underline{x}-\underline{z}))d\underline{x} \; ,$$

$$= \left((2\pi)^{-n} \int_{\underline{R}_n} (1+|\eta\underline{y}|)^{\omega} \alpha^{\wedge}(\underline{y})\, d\underline{y}\right) |||u^{\wedge}|||^2 \ .$$

Since $\alpha^{\wedge}(\underline{y}) = O(|\underline{y}|^{-s})$ as $\underline{y} \to \infty$ for every s, a simple application of the Lebesgue limit theorem yields

$$\overline{\lim_{\eta \to 0}}\ (2\pi)^{-n} \int_{\underline{R}_n} (1+|\eta\underline{y}|)^{\omega} \alpha^{\wedge}(\underline{y})\, d\underline{y} = 1 \ .$$

Therefore we have

(4) $$\overline{\lim_{\eta \to 0}}\ |||\alpha^{\wedge}_{\eta} * u^{\wedge}_m|||^2 \leq |||u^{\wedge}_m|||^2 \ .$$

We assert that $\alpha^{\wedge}_{\eta} * u^{\wedge}_m \rightharpoonup u^{\wedge}_m$ as $\eta \to 0$ in $\underset{\sim}{K}^{\wedge}$. Let $v^{\wedge} \in \underset{\sim}{H}^{\wedge}$ be bounded with bounded support. Then we have

$$\lim_{\eta \to 0} (\alpha^{\wedge}_{\eta} * u^{\wedge} | v^{\wedge}) = \lim_{\eta \to 0} (2\pi)^{-n} \int_{\underline{R}_n} (1+|\underline{x}|)^{\omega} (\alpha^{\wedge}_{\eta} * u^{\wedge}_m)(\underline{x}) \overline{v^{\wedge}(\underline{x})}\, dx \ .$$

By an elementary fact concerning approximate identities

$$\alpha^{\wedge}_{\eta} * u^{\wedge}_m \to u^{\wedge}_m \quad \text{in} \quad \underset{\sim}{H}^{\wedge} \quad \text{as} \quad \eta \to 0 \ .$$

Since $(1+|\underline{x}|)^{\omega}\, v^{\wedge}(\underline{x}) \in \underset{\sim}{H}^{\wedge}$ we then have

$$\lim_{\eta \to 0} (\alpha^{\wedge}_{\eta} * u^{\wedge}_m | v^{\wedge}) = (2\pi)^{-n} \int_{\underline{R}_n} (1+|\underline{x}|)^{\omega}\, u^{\wedge}_m(\underline{x}) \overline{v^{\wedge}(\underline{x})}\, d\underline{x} = (u^{\wedge} | v^{\wedge}).$$

Since functions which are bounded and have bounded support are dense in $\underset{\sim}{K}^{\wedge}$ and since (4) holds we have

(5) $$\alpha^{\wedge}_{\eta} * u^{\wedge}_m \rightharpoonup u^{\wedge}_m \quad \text{in} \quad \underset{\sim}{K}^{\wedge}.$$

Note that (4) and (5) together imply that

$$\lim_{\eta \to 0} |||\alpha_\eta^{\wedge} * u_m^{\wedge} - u_m^{\wedge}||| = 0 \quad ,$$

and that the function $\alpha_\eta(\underline{y})u_m(\underline{y}) = \alpha(\eta\underline{y})u_m(\underline{y})$ is supported in $\Omega \cap \{\underline{y}:|\underline{y}| \leq \eta^{-1}\}$. It follows that if we set $v_k^{\wedge} = \alpha_\eta^{\wedge} * u_k^{\wedge}$ $k = 1,\ldots,n$ where η is sufficiently small then $\underline{v}^{\wedge}$ will have the required properties.

b. Suppose now that $\underline{v}^{\wedge} \in K_{\underline{M}}^{\wedge}$ and that $v_k(\underline{y}) = 0$ for $\underline{y} \notin \Omega_b$ $k = 1,\ldots,p$ where Ω_b is a bounded subset of Ω . We assert that $\underline{v}^{\wedge}(\underline{x})$ can be represented as a sum

$$\underline{v}^{\wedge}(\underline{x}) = \sum_{r=1}^{R} \underline{v}^{\wedge}(r,\underline{x})$$

where $\underline{v}^{\wedge}(r,\underline{x}) \in K_{\underline{M}}^{\wedge}$ $r = 1,\ldots,R$ and corresponding to each r there is a bounded star-shaped subset Ω_r of Ω such that for $k = 1,\ldots,p$

$$v_k^{\wedge}(r,\underline{y}) = 0 \qquad \text{for} \qquad \underline{y} \notin \Omega_r \qquad r = 1,\ldots,R \ .$$

Let $U_1,\ldots,U_R$ be bounded open sets covering $\overline{\Omega_b}$ such that each $\Omega \cap U_r$ is star-shaped. Let β_r $r = 1,\ldots,R$ be C^∞ functions such that β_r is supported in U_r and $\sum_1^R \beta_r(\underline{y}) = 1$ on Ω_b . If we set

$$v^{\wedge}(r,\underline{y}) = \beta_r(\underline{y})\underline{v}^{\wedge}(\underline{y})$$

then the $\underline{v}^{\wedge}(r,\underline{y})$ $r = 1,\ldots,R$ have the desired properties. We need only check that $\underline{v}^{\wedge}(r,y) \in \underset{\sim}{K}^{\wedge}$. We have

$$|||\underline{v}_k^{\wedge}(r,\underline{y})|||^2 = (2\pi)^{-n}\int_{\underline{R}_n}(1+|\underline{x}|)^{\omega}\,d\underline{x}\left|(2\pi)^{-n}\int_{\underline{R}_n}\beta_r^{\wedge}(\underline{x}-\underline{z})v_k^{\wedge}(\underline{z})\,d\underline{z}\right|^2$$

$$\leq (2\pi)^{-n}\left(\int_{\underline{R}_n}(1+|\underline{x}|)^{\omega}\,d\underline{x}\,(2\pi)^{-n}\int_{\underline{R}_n}|\beta_r^{\wedge}(\underline{x}-\underline{z})|\cdot|v_k^{\wedge}(\underline{z})|^2 d\underline{z}\right)$$

$$\times\left\{(2\pi)^{-n}\int_{\underline{R}_n}|\beta_r^{\wedge}(\underline{x}-\underline{z})|\,d\underline{z}\right\} \quad .$$

Since $(1+|\underline{x}|)^{\omega}$ is submultiplicative we have

$$|||\underline{v}_k^{\wedge}(r,\underline{y})|||^2 \leq \left\{(2\pi)^{-n}\int_{\underline{R}_n}|\beta_r^{\wedge}(\underline{x}-\underline{z})|\,d\underline{z}\right) \times \quad (2\pi)^{-n}\int_{\underline{R}_n}(1+|\underline{y}|)^{\omega}|v_k^{\wedge}(r,\underline{y})|^2 d\underline{y}\right\}$$

$$\times\left\{(2\pi)^{-n}\int_{\underline{R}_n}(1+|\underline{x}-\underline{y}|)^{\omega}\,\beta_r^{\wedge}(\underline{x}-\underline{y})\,d\underline{x}\right\} \quad ,$$

$$\leq A|||v_k^{\wedge}|||^2 \quad .$$

Proof of Part 2. We are given $\underline{u}^{\wedge} \in \underset{\sim}{K}$ such that $u_k(\underline{y}) = 0$ for $y \notin \Omega_s$ $k = 1,\ldots,p$ where Ω_s is a bounded star-shaped subset of Ω. Let Ω_s be star-shaped with respect to $\underline{y}_o$. We set

$$\underline{v}(\underline{y}) = \underline{u}[\underline{y}_o + r^{-1}(\underline{y}-\underline{y}_o)]$$

where r, $0 < r < 1$, will be fixed shortly. It is clear that $\underline{v}(\underline{y})$ vanishes outside $r(\Omega-\underline{y}_o) + \underline{y}_o$. A short computation shows that

$$\underline{v}^{\wedge}(\underline{x}) = r^{n}\underline{u}^{\wedge}(r\underline{x})\exp[i(1-r)\underline{x}\cdot\underline{y}_{o}] .$$

Clearly we can choose r so close to 1 that

$$\|\underline{u}^{\wedge} - \underline{v}^{\wedge}\|_{\underset{\sim}{H}^{\wedge}} < \frac{\epsilon}{2} , \quad |||\underline{u}^{\wedge} - \underline{v}^{\wedge}||| < \frac{\epsilon}{2} .$$

Let $\epsilon_{r} = \operatorname{dist}\{r(\Omega - \underline{y}_{o}) + \underline{y}_{o} , \Omega'\}$ where Ω' is the complement of Ω. Let $\alpha(\underline{y})$ be a non-negative C^{∞} function on $\underline{R}_{n}$ supported in $|\underline{y}| \leq 1$ and such that

$$\int_{\underline{R}_{n}} \alpha(\underline{y})d\underline{y} = 1 ,$$

and set

$$w(\underline{y}) = \eta^{-n}\int_{\underline{R}_{n}} \alpha[\eta^{-1}(\underline{y}-\underline{z})]v(\underline{z})d\underline{z}$$

where $0 < \eta < \epsilon_{r}$. Then $\underline{w} \in \underset{\sim}{D}(\Omega)$ and

$$\underline{w}^{\wedge}(\underline{x}) = \alpha^{\wedge}(\eta\underline{x})\underline{w}^{\wedge}(\underline{x}) .$$

Since $\alpha^{\wedge}(0) = 1$ and $|\alpha^{\wedge}(\underline{x})| = 1$ it is more or less obvious that

$$\lim_{\eta \to 0+} \|\underline{v}^{\wedge} - \underline{w}^{\wedge}\|_{\underset{\sim}{H}^{\wedge}} = \lim_{\eta \to 0+} |||\underline{v}^{\wedge} - \underline{w}^{\wedge}||| = 0 .$$

Choose $\eta > 0$ so small that

$$\|\underline{v}^{\wedge} - \underline{w}^{\wedge}\|_{\underset{\sim}{H}^{\wedge}} < \frac{\epsilon}{2} , \quad |||\underline{v}^{\wedge} - \underline{w}^{\wedge}||| < \frac{\epsilon}{2} .$$

Our proof is now complete

Theorem 6b. $(\underline{S}^{\wedge})^{\frac{1}{2}}$ is the closure of the strong limit of $(\underline{S}^{\wedge}(t))^{\frac{1}{2}}\,\underline{F}^{\wedge}(t)$ on $\underset{\sim}{M}$ as $t \to +\infty$.

Proof. This follows almost immediately from Theorems 5a, 5b, 5c and 6a.

7. The Asymptotic Formula, I

Let $\underline{S}^{\wedge}_{\underline{F}}$ be constructed from $\underline{F}^{\wedge}$ and $\underline{S}^{\wedge}$ as in Section 1 of Chapter II. Let

$$\underline{S}^{\wedge}_{\underline{F}} = \int_{0-}^{\infty} \lambda \, d\, \underline{E}^{\wedge}(\lambda)$$

be the spectral resolution of $\underline{S}^{\wedge}_{\underline{F}}$ on $\underline{M}^{\wedge} = \underline{F}^{\wedge}\underline{H}^{\wedge}$, and let

$$\underline{S}^{\wedge}_{\underline{F}}(t) = \int_{0-}^{\infty} \lambda \, d\, \underline{E}^{\wedge}_t(\lambda)$$

be the spectral resolution of $\underline{S}^{\wedge}_{\underline{F}}(t) = \underline{F}^{\wedge}(t)\underline{S}^{\wedge}(t)\underline{F}^{\wedge}(t)$ on $\underline{H}^{\wedge}$.

It follows from Sections 1-4 and Corollary 4b of Chapter II that

$$(1) \qquad \underline{E}^{\wedge}_t(\lambda) \to \underline{E}^{\wedge}(\lambda) \qquad 0 \leq \lambda < \infty$$

for every $\lambda \notin \sigma_p(\underline{S}^{\wedge}_{\underline{F}})$.

Lemma 7a. For $0 < t < \infty$ let $\{\underline{u}^{\wedge}(t,\underline{x})\}$ be a family of functions such that

i. $\underline{u}^{\wedge}(t,\underline{x}) \in \underline{F}^{\wedge}(t)\underline{H}^{\wedge}$, for all $t > 0$

ii. $\|\underline{u}^{\wedge}(t,\underline{x})\|_{\underline{H}^{\wedge}} = 1$, for all $t > 0$.

Then for each k, $1 \leq k \leq p$, $\{u^{\wedge}_k(t,\underline{x})\}$ is a uniformly bounded and equicontinuous family of functions on R^t_k.

Proof. Since $\underline{u}^{\wedge}(t,\underline{x}) \in \underline{F}^{\wedge}(t)\,\underline{H}^{\wedge}$ we have $\underline{u}^{\wedge}(t) = \underline{F}^{\wedge}(t)\underline{u}^{\wedge}(t)$.

Hence

$$(2) \qquad u_k^\wedge(t,\underline{x}) = \sum_{m=1}^{p} F^\wedge(t,k,m)u_m^\wedge(t,\underline{x}), \qquad k = 1,\dots,p \quad ,$$

where

$$F^\wedge(t,k,m)u_m^\wedge(t,\underline{x}) = \begin{cases} \int_\Omega e^{i\underline{y}\cdot[\underline{x}+t(\underline{\xi}_k-\underline{\xi}_m)]} u_m(t,\underline{y})d\underline{y} \ , & \underline{x} \in R_k^t \\ 0 & x \notin R_k^t \end{cases} .$$

We have

$$\|u_m(t,\underline{y})\|_{\underset{\sim}{H}} = \|u_m^\wedge(t,\underline{x})\|_{\underset{\sim}{H}^\wedge} \le \|\underline{u}^\wedge(\underline{x},t)\|_{\underset{\approx}{H}^\wedge} = 1 \quad .$$

Thus by Schwarz's inequality

$$|F^\wedge(t,k,m)u_m^\wedge(t,\underline{x})|^2 \le \int_\Omega d\underline{y} \cdot \int_\Omega |u_m(t,\underline{y})|^2 \, d\underline{y} \le \int_\Omega d\underline{y} \quad .$$

In view of (2) this proves the $\{u^\wedge(t,\underline{x})\}$ are uniformly bounded.

We now show that $\{\underline{u}_k^\wedge(t,\underline{x})\}$ is an equicontinuous family on R_k^t. From (2) it is clear that this will follow if we show for fixed ℓ and m that

$$\{F^\wedge(t,\ell,m)u_m^\wedge(t,\underline{x})\}_t$$

is an equicontinuous family on R_k^t.

Let $\epsilon > 0$ be given. Partition Ω into disjoint subsets Ω_1 and Ω_2 where Ω_1 is bounded and $|\Omega_2| < \epsilon^2/16$ and then write

$$F^\wedge(t,k,m)u_m^\wedge(t,\underline{x}) = v^\wedge(t,\underline{x}) + w^\wedge(t,\underline{x})$$

where

$$v^\wedge(t,\underline{x}) = \int_{\Omega_1} e^{i\underline{y}\cdot[\underline{x}+t(\underline{\xi}_k-\underline{\xi}_m)]} u_m(t,\underline{y})d\underline{y}$$

$$w^\wedge(t,\underline{x}) = \int_{\Omega_2} e^{i\underline{y}\cdot[\underline{x}+t(\underline{\xi}_k-\underline{\xi}_m)]} u_m(t,\underline{y})d\underline{y} \quad .$$

By Schwarz's inequality

$$|w^\wedge(t,\underline{x})| \leq |\Omega_2|^{\frac{1}{2}} \|u_m(t,\underline{y})\|_{\underset{\sim}{H}} < \epsilon/4 \quad .$$

On the other hand, since if $\underline{y} \in \Omega_1$

$$|e^{i\underline{y}\cdot\underline{x}} - e^{i\underline{y}\cdot\underline{x}'}| \leq \rho|\underline{x}-\underline{x}'|$$

where $\rho = \sup\{|\underline{y}| : \underline{y} \in \Omega_1\}$ it follows that

$$|v^\wedge(t,\underline{x}) - v^\wedge(t,\underline{x}')| \leq K|\underline{x}-\underline{x}'|$$

where $K = \rho|\Omega_1|^{\frac{1}{2}}$. Thus

$$|F^\wedge(t,\ell,m)u_m^\wedge(t,\underline{x}) - F^\wedge(t,\ell,m)u_m^\wedge(t,\underline{x}')|$$

$$\leq |v^\wedge(t,\underline{x}) - v^\wedge(\underline{t},\underline{x}')| + |w^\wedge(t,\underline{x})| + |w^\wedge(t,\underline{x}')| \leq K|\underline{x}-\underline{x}'| + \frac{\epsilon}{4} + \frac{\epsilon}{4} ,$$

etc.

Lemma 7b. Let $\{\underline{u}^\wedge(t,\underline{x})\}$, $t > 0$, be a family of functions such that

i. $\underline{u}^\wedge(t,\underline{x}) \in \underline{F}^\wedge(t)\underset{\approx}{H}^\wedge$.

ii. $\|\underline{u}^\wedge(t)\|_{\underset{\approx}{H}^\wedge} = 1$.

iii. $$\langle \underline{S}^{\wedge}_{\underline{F}}(t)\underline{u}^{\wedge}(t)|\underline{u}^{\wedge}(t)\rangle_{\underset{\approx}{H}^{\wedge}} \leq M < \infty \quad .$$

We assert that if $\underline{u}^{\wedge}(t) \longrightarrow \underline{u}^{\wedge}$ as $t \to \infty$ in $\underset{\sim}{P}_1$, a subsequence of $\underset{\sim}{P}$, then $\underline{u}^{\wedge} \not\equiv 0$.

Proof. By Lemma 7a for $m = 1,\ldots,p$, $\{u^{\wedge}_m(t,\underline{x})\}$, $t > 0$, is a uniformly bounded and equicontinuous family of functions on R^t_m. By Arzela's theorem there exists a subsequence $\underset{\sim}{P}_2$ of $\underset{\sim}{P}_1$ such that $\underline{u}^{\wedge}(t,\underline{x})$ converges uniformly on any compact subset of $\underline{R}_n$ to a function which necessarily is $\underline{u}^{\wedge}(x)$. In particular

(3) $$\lim_{\substack{t \to \infty \\ t \in \underset{\sim}{P}_2}} u^{\wedge}_m(t,\underline{x}) = u_m(\underline{x}) \qquad m = 1,\ldots,p \quad ,$$

uniformly for $\underline{x} \in N(A) = \{\underline{x} : |\underline{x}| \leq A\}$ for any $A > 0$. Given any $M_1 > 0$ there exists $A_o > 0$ and $t_o > 0$ such that if $t \geq t_o$

$$g^t_m(\underline{x}) \geq M_1$$

for all $\underline{x} \in R^t_m \backslash N(A_o)$, $m = 1,\ldots,p$. We have

$$\sum_{m=1}^{p} \int_{N(A_o)} g^t_m(\underline{x})|u^{\wedge}_m(t,\underline{x})|^2 \, d\underline{x} + \sum_{m=1}^{p} \int_{\underline{R}_n \backslash N(A_o)} g^t_m(\underline{x})|u^{\wedge}_m(t,\underline{x})|^2 \, d\underline{x}$$

$$= \langle \underline{S}^{\wedge}(t,\underline{F}^{\wedge})\underline{u}^{\wedge}(t)|\underline{u}^{\wedge}(t)\rangle_{\underset{\approx}{H}^{\wedge}} \leq M < \infty \quad .$$

Therefore

$$\sum_{m=1}^{p} \int_{\underline{R}_n \backslash N(A_o)} g^t_m(\underline{x})|u^{\wedge}_m(t,\underline{x})|^2 \, d\underline{x} \leq M \quad .$$

If $t \geq t_o$

$$\sum_{m=1}^{p} \int_{\underline{R}_n \setminus N(A_o)} g_m^t(\underline{x}) |u_m^\wedge(t,\underline{x})|^2 \, d\underline{x} \geq M_1 \sum_{m=1}^{p} \int_{\underline{R}_n \setminus N(A_o)} |u_m^\wedge(t,\underline{x})|^2 \, d\underline{x} \quad .$$

Thus if $t \geq t_o$

$$M_1 \sum_{m=1}^{p} \int_{\underline{R}_n \setminus N(A_o)} |u_m^\wedge(t,\underline{x})|^2 \, d\underline{x} \leq M \quad ,$$

and thus, since $\|\underline{u}^\wedge(t)\|_{\underline{H}^\wedge} = 1$,

(4)
$$\sum_{m=1}^{p} \int_{N(A_o)} |\underline{u}^\wedge(t,\underline{x})|^2 \, d\underline{x} \geq 1 - M/M_1 \quad .$$

Since (3) holds, the inequality (4) implies that

$$\sum_{m=1}^{p} \int_{N(A_o)} |u_m^\wedge(\underline{x})|^2 \, d\underline{x} \geq 1 - M/M_1 > 0$$

and thus $\underline{u} \neq \underline{0}$ as desired.

It follows from Lemma 7b and Theorem 5a of I that the spectrum of $\underline{S}_{\underline{F}^\wedge}^\wedge$ lies in $(0,\infty)$ and that for each $\lambda > 0$ the part or the spectrum in $(0,\lambda)$ consists of finitely many eigen values each having finite multiplicity. Since

$$\underline{S}_{\underline{F}}^\wedge = S_F^\wedge(1) \oplus \cdots \oplus S_F^\wedge(p) \quad ,$$

the spectrum of each $S_F^\wedge(m)$ has the same character. Let

(5)
$$0 < \mu_m(1) \leq \mu_m(2) \leq \cdots, \quad \lim_{k \to \infty} \mu_m(k) = \infty$$

be the eigen values of $S_F^{\wedge}(m)$, each repeated according to its multiplicity. Note that the eigen values (5) are completely determined by the triple

(6) $$(\Omega,\omega,\Phi_m(\underline{x})) \quad .$$

It is clear the $\{\mu(k)\}_1^{\infty}$ the set of eigen values of $\underline{S}_F^{\wedge}$ is

$$\{\mu_1(k)\}_1^{\infty} \cup \cdots \cup \{\mu_p(k)\}_1^{\infty} \quad ,$$

written in non-decreasing order.

Theorem 7c. Under assumptions i.- v. of §1, the bottom of the spectrum of $T^{\wedge}(t)$ becomes discretized as $t \to +\infty$ and

(7) $$\lim_{t\to\infty} t^{\omega} L(t^{-1})^{-1} \lambda(k,t) = \mu(k) \qquad k= 1,2,\ldots$$

where $0 < \lambda(1,t) \le \lambda(2,5) \le \cdots \le \lambda(k,t)$ are for t sufficiently large the lowest eigen values of $T^{\wedge}(t)$ repeated according to their multiplicities.

Proof. This follows from the perturbation theory of Chapter II the assumptions of which we have now verified.

8. The Asymptotic Formula, II

We are now ready to prove a result which is more general then Theorem 7c in that we suppose that, in addition to q zeros like those previously considered, $f(\underline{\xi})$ has p-q zeros of lower order. As we might anticipate the zeros of lower order do not effect the final result. Let us replace iv. and v. of §1 by the following.

iv'. There exists an integer q, $1 \le q \le p$ and q non-negative measurable functions $\Phi_m(\underline{\xi})$ which are homogeneous of degree 0 on $\underline{R}_n$ and which are bounded and bounded away from 0.

v'. For each $m = 1,\ldots,q$ given $\epsilon > 0$ there exists a $\delta > 0$ such that $0 < |\underline{\xi} - \underline{\xi}_m| < \delta$ implies that

$$\left| \frac{f(\underline{\xi})}{L(|\underline{\xi} - \underline{\xi}_m|)|\underline{\xi} - \underline{\xi}_m|^{\omega}} - \Phi_m(\underline{\xi} - \underline{\xi}_m)\right| < \epsilon \quad .$$

For each $m = q+1,\ldots,p$ and each $T > 0$ there exists a $\delta > 0$ such that $0 < |\underline{\xi} - \underline{\xi}_m| \le \delta$ implies that

$$\frac{f(\underline{\xi})}{L(|\underline{\xi} - \underline{\xi}_m|)|\underline{\xi} - \underline{\xi}|^{\omega}} \ge T \quad .$$

Let,

$$0 < \lambda(1,t) \le \lambda(2,t) \le \cdots$$

be as in the previous section and let $\{\mu(k)\}_1^\infty$ be the non-decreasing rearrangement of

$$\{\mu_1(k)\}_1^\infty \cup \cdots \cup \{\mu_q(k)\}_1^\infty \quad ,$$

where $\{\mu_m(k)\}_1^\infty$ are the eigen values of $S(m)\hat{}_{F}\hat{}$ and therefore are determined by $\{\Omega,\omega,\Phi_m\}$.

Theorem 8a. Under assumptions i. -- iii. of §1, iv', and v' we have

$$\lim_{t\to\infty} t^{\omega}L(t^{-1})^{-1}\,\lambda(k,t) = \mu(k) \qquad k = 1,2,\ldots \quad .$$

Proof. For $T > 0$ fixed let us define two new functions, $f*(\underline{\xi})$

and $f_*(\underline{\xi})$ as follows:

$$f^*(\underline{\xi}) = \begin{cases} 1 & \text{if } |\underline{\xi} - \underline{\xi}_m| \leq \delta \quad m = q+1,\dots,p \\ f(\underline{\xi}) & \text{all other } \underline{\xi} \end{cases},$$

$$f_*(\underline{\xi}) = \begin{cases} TL(|\underline{\xi}-\underline{\xi}_m|)|\xi - \xi_m|^{\omega} & \text{if } |\underline{\xi} - \underline{\xi}_m| \leq \delta \quad m = q \quad 1,\dots,p \\ f(\underline{\xi}) & \text{all other } \underline{\xi} \end{cases}.$$

If we choose δ, depending on T, sufficiently small then

(1) $$f_*(\underline{\xi}) \leq f(\underline{\xi}) \leq f^*(\underline{\xi}) .$$

Both f_* and f^* satisfy assumptions i. - v. of §1. By Theorem 7c applied to f_*

(2) $$\lim_{t \to \infty} t^{\omega} L(t^{-1})^{-1} \lambda(t,k,f^*) = \mu(k) \qquad k = 1,2,\dots$$

where $\{\mu(k)\}_1^{\infty}$ is defined above. Let $0 < \nu(1) \leq \nu(2) \leq \cdots$ be the eigen values associated with the triple $\{\Omega,\omega,1\}$ and let $\mu(T,k)$ $k = 1,2,\cdots$ be the numbers

$$\{\mu_1(k)\}_1^{\infty} \cup \cdots \cup \{\mu_q(k)\}_1^{\infty} \cup \{T\nu(k)\}_1^{\infty} \cup \cdots \cup \{T\nu(k)\}_1^{\infty}$$

(there are altogether p sets) arranged in non-decreasing order. By Theorem 7c applied to f_*

(3) $$\lim_{t \to \infty} t^{\omega} L(t^{-1})^{-1} \lambda(f_*,k,t) = \mu(T,k) \qquad k = 1,2,\cdots .$$

By (1) and the Weyl-Courant lemma

(4) $$\lambda(f_*,k,t) \leq \lambda(f,k,t) \leq \lambda(f^*,k,t) \quad k = 1,2,\cdots \quad .$$

It is evident that for k fixed

(5) $$\mu(T,k) = \mu(k)$$

provided T is sufficiently large. Combining (2) -- (5) we obtain our desired result.

Chapter IV

THE FOURIER SERIES THEOREM

1. Spaces and Operators

This chpater is devoted to the discrete analogue of the theory of Chapter III. We begin by making a series of assumptions analogous to those made in §1 of III.

i. We are given a set Ω in $\underline{R}_n$ of positive measure which is bounded and locally star-shaped.

ii. We have a real function $f(\underline{\theta})$ on $\underline{T}_n$ which is bounded and measurable and which assumes its minimum 0 at p distinct points $\underline{\theta}_m$ $m = 1,\dots,p$ and is such that

$$\inf\{f(\underline{\theta}):|\underline{\theta}-\underline{\theta}_m| \geq \delta\} \quad m = 1,\dots,p\} > 0$$

for each $\delta > 0$.

iii. This is the same as assumption iii. of §1, III.

iv. This is the same as assumption iv. of §1, III.

v. For each $m = 1,\dots,p$ given $\epsilon > 0$ there exists $\delta > 0$ such that $0 < |\underline{\theta}-\underline{\theta}_m| < \delta$ implies that

$$\left| \frac{f(\underline{\theta})}{L(|\underline{\theta}-\underline{\theta}_m|)|\underline{\theta}-\underline{\theta}_m|^{\omega}} - \Phi_m(\underline{\theta}-\underline{\theta}_m) \right| < \epsilon \quad .$$

Let $E_{\mathcal{M}}$ be the Hilbert space of complex functions defined on $\underline{Z}_n$ (the n-dimensional lattice group) defined by the inner product

$$\langle u|v\rangle_{E_{\mathcal{M}}} = \sum_{\underline{Z}_n} u(\underline{k})\overline{v(\underline{k})}$$

where $\underline{k} = (k_1,\ldots,k_n)$. Let $\underset{\sim}{E}^{\wedge}$ be the Hilbert space of complex Lebesgue measurable functions on $\underline{T}_n$ defined by the inner product

$$\langle u^{\wedge} | v^{\wedge} \rangle_{\underset{\sim}{E}^{\wedge}} = (2\pi)^{-n} \int_{\underline{T}_n} u^{\wedge}(\underline{\theta})\overline{v^{\wedge}(\underline{\theta})}d\underline{\theta} \,,$$

where $\underline{\theta} = (\theta_1,\ldots,\theta_n)$, etc. The mapping φ from $\underset{\sim}{E}$ to $\underset{\sim}{E}^{\wedge}$ defined by

$$\varphi u \cdot (\underline{\theta}) = \sum_{\underline{Z}_n} u(\underline{k})e^{i\underline{k}\cdot\underline{\theta}}$$

is unitary. Its inverse φ^{-1} is given by

$$\varphi^{-1}u^{\wedge} \cdot (\underline{k}) = (2\pi)^{-n} \int_{\underline{T}_n} u^{\wedge}(\underline{\theta})e^{-i\underline{k}\cdot\underline{\theta}}d\underline{\theta} \;.$$

Here $\underline{k}\cdot\underline{\theta} = k_1\theta_1 + \cdots + k_n\theta_n$.

We set $\Omega_t = t\,\Omega \cap \underline{Z}_n$, $0 < t < \infty$, and define

$$E(t)u \cdot (\underline{k}) = \begin{cases} u(\underline{k}) & \underline{k} \in \Omega_t \\ 0 & \underline{k} \notin \Omega_t \end{cases}$$

$E^{\wedge}(t)$ on $\underset{\sim}{E}^{\wedge}$ is defined by

$$E^{\wedge}(t) = \varphi\, E(t)\varphi^{-1} .$$

$E(t)$ and $E^{\wedge}(t)$ are projections.

We define $T^{\wedge}$ on $\underset{\sim}{E}^{\wedge}$ by

$$T^{\wedge}u^{\wedge}(\underline{\theta}) = f(\underline{\theta})u^{\wedge}(\theta)$$

and consider

$$E^{\wedge}(t)T^{\wedge}E^{\wedge}(t)\Big|_{E^{\wedge}(t)\underset{\mathcal{M}}{E^{\wedge}}} = T^{\wedge}(t) \quad .$$

The object of the present chapter is to determine the asymptotic behaviour of the eigen values of the operator $T^{\wedge}(t)$.

The spaces $\underset{\mathcal{M}}{H}$ and $\underset{\mathcal{M}}{H^{\wedge}}$ will be the same as in Chapter III. We recall that $\underset{\mathcal{M}}{H}$ and $\underset{\mathcal{M}}{H^{\wedge}}$ are as Hilbert spaces of complex measurable functions defined in $\underline{R}_n$ with inner products

$$\langle u|v\rangle_{\underset{\mathcal{M}}{H}} = \int_{\underline{R}_n} u(\underline{y})\overline{v(\underline{y})}d\underline{y} \quad ,$$

and

$$\langle u^{\wedge}|v^{\wedge}\rangle_{\underset{\mathcal{M}}{H^{\wedge}}} = (2\pi)^{-n}\int_{\underline{R}_n} u^{\wedge}(\underline{x})\overline{v^{\wedge}(\underline{x})}d\underline{x} \quad .$$

The mapping $\psi : H \rightarrow H^{\wedge}$ defined by $\psi : u \rightarrow u^{\wedge}$ where

$$\psi u(\underline{x}) = \int_{\underline{R}_n} u(\underline{y})e^{i\underline{x}\cdot\underline{y}}d\underline{y}$$

and its inverse $\psi^{-1} : \underset{\mathcal{M}}{H^{\wedge}} \rightarrow \underset{\mathcal{M}}{H^{\wedge}}$ defined by

$$\psi^{-1}u^{\wedge}(\underline{y}) = (2\pi)^{-n}\int_{\underline{R}_n} u^{\wedge}(\underline{x})e^{-i\underline{x}\cdot\underline{y}}d\underline{y}$$

are unitary mappings. As indicated in these definitions we will reserve the variables $\underline{y}$, $\underline{x}$ for elements of $\underset{\mathcal{M}}{H}$, $\underset{\mathcal{M}}{H^{\wedge}}$ respectively.

$\underline{\underset{\mathcal{M}}{H}}$ and $\underline{\underset{\mathcal{M}}{H^{\wedge}}}$ consist of elements which are ordered p-tuples of functions in $\underset{\mathcal{M}}{H}$ and $\underset{\mathcal{M}}{H^{\wedge}}$, respectively. The inner products in $\underline{\underset{\mathcal{M}}{H}}$ and $\underline{\underset{\mathcal{M}}{H^{\wedge}}}$ are defined by

$$\langle \underline{u} | \underline{v} \rangle_{\underset{\sim}{H}} = \langle [u_1,\ldots,u_p] | [v_1,\ldots,v_p] \rangle = \sum_{m=1}^{p} \langle u_m | v_m \rangle_{\underset{\sim}{H}}$$

and

$$\langle \underline{u}^{\wedge} | \underline{v}^{\wedge} \rangle_{\underset{\sim}{H}^{\wedge}} = \langle [u_1^{\wedge},\ldots,u_p^{\wedge}] | [v_1^{\wedge},\ldots,v_p^{\wedge}] \rangle_{\underset{\sim}{H}^{\wedge}} = \sum_{m=1}^{p} \langle u_m^{\wedge} | v_m^{\wedge} \rangle_{\underset{\sim}{H}^{\wedge}} .$$

Let $\underline{T}_n$ be represented in $\underline{R}_n$ by the set $\{\underline{\theta}: -\pi \leq \theta_j < \pi,\ 1 \leq j \leq n\}$. We may assume without loss of generality that the set just described can be partitioned into p non-overlapping rectanges $R_1,\ldots,R_p$ such that $\underline{\theta}_m$ lies in the interior of R_m $m = 1,\ldots,p$.

For each $t > 0$, $m = 1,\ldots,p$ we define a mapping $\sigma(t,m)$ of R_m into $\underline{R}_n$ by

$$\underline{x} = \sigma(t,m)(\underline{\theta}) = t(\underline{\theta} - \underline{\theta}_m) .$$

We denote by R_m^t the image of R_m under this mapping. Similarly the inverse of $\sigma(t,m)$, denoted $\tau(t,m)$, is defined by

$$\underline{\theta} = \tau(t,m)(\underline{x}) = t^{-1}\underline{x} + \underline{\theta}_m .$$

For each $t > 0$, $m = 1,\ldots,p$ we define a mapping $\chi(t,m)$ from $\underset{\sim}{E}^{\wedge}$ to $\underset{\sim}{H}^{\wedge}$ as follows:

if $u^{\wedge}(\underline{\theta}) \in \underset{\sim}{E}^{\wedge}$

$$\chi(t,m)u^{\wedge}.(\underline{x}) = \begin{cases} t^{-n/2} u^{\wedge} \circ \tau(t,m)(\underline{x}) & \underline{x} \in R_m^t \\ 0 & \underline{x} \notin R_m^t \end{cases} .$$

Next we define $\underline{\chi}(t) : \underset{\sim}{E}^{\wedge} \to \underset{\sim}{H}^{\wedge}$ by

$$\underline{\chi}(t)u^{\wedge}(\underline{x}) = [\chi(t,1)u^{\wedge}(\underline{x}),\ldots,\chi(t,p)u^{\wedge}(\underline{x})] \quad .$$

As in Section 1 of III $\underline{\chi}(t)$ is an isometry.

The adjoint of $\underline{\chi}(t)$ is defined as follows. We first define $\chi^*(t,m): \underset{\mathcal{M}}{H^{\wedge}} \to \underset{\mathcal{M}}{E^{\wedge}}$, for each $t > 0$, $m = 1,\ldots,p$ by

$$\chi^*(t,m)u^{\wedge}(\underline{\theta}) = \begin{cases} t^{n/2}u^{\wedge}\circ\sigma(t,m)(\underline{\theta}) & \underline{\theta} \in R_m^t \\ 0 & \underline{\theta} \notin R_m^t \end{cases} \quad .$$

If $\underline{u}^{\wedge} = [u_1^{\wedge},\ldots,u_p^{\wedge}] \in \underset{\underline{\mathcal{M}}}{H^{\wedge}}$ then $\underline{\chi}^*(t)\underline{u}(\underline{\theta})$, $\underline{\theta} \in \underline{T}_n$ is defined by

$$\underline{\chi}^*(t)\underline{u}^{\wedge}(\underline{\theta}) = \begin{cases} \chi^*(t,1)u_1^{\wedge}(\underline{\theta}) & \underline{\theta} \in R_1 \\ \vdots & \vdots \\ \chi^*(t,p)u_p^{\wedge}(\underline{\theta}) & \underline{\theta} \in R_p \end{cases}$$

This function is well-defined at each $\underline{\theta}$ since the R_m 's are a disjoint partition of $\underline{T}_n$. As in III we have

$$\underline{\chi}^*(t)\underline{\chi}(t) = I$$

$$\underline{\chi}(t)\underline{\chi}^*(t) = \begin{cases} I & \text{on } \underline{\chi}(t)\underset{\mathcal{M}}{E^{\wedge}} \\ 0 & \text{on } (\underline{\chi}(t)\underset{\mathcal{M}}{E^{\wedge}})^{\perp} \end{cases} \quad .$$

2. Application of the Pertubation Theory

In this section we introduce various operators on the Hilbert spaces defined in section 1.

a) We recall that

$$E(t)u(\underline{k}) = \begin{cases} u(\underline{k}) & \underline{k} \in \Omega_t \\ 0 & \underline{k} \notin \Omega_t \end{cases} ,$$

and that $E^{\wedge}(t)$ is defined on $\underset{\sim}{E}^{\wedge}$ by $E^{\wedge}(t) = \varphi E(t)\varphi^{-1}$. This implies that

$$E^{\wedge}(t)u^{\wedge}(\underline{\theta}) = \sum_{\underline{k}\in\Omega_t} u(\underline{k})e^{i\underline{k}\cdot\underline{\theta}}$$

where

$$u^{\wedge}(\underline{\theta}) = \sum_{\underline{Z}_n} u(\underline{k})e^{i\underline{k}\cdot\underline{\theta}}$$

$\underline{F}^{\wedge}(t)$ on $\underset{\approx}{H}^{\wedge}$ is defined by $\underline{F}^{\wedge}(t) = \underline{\chi}(t)E^{\wedge}(t)\underline{\chi}^{*}(t)$. We note that

$$\begin{aligned} &E(t) \text{ is a projection on } \underset{\sim}{E},\\ &E^{\wedge}(t) \text{ is a projection on } \underset{\sim}{E}^{\wedge},\\ &\underline{F}^{\wedge}(t) \text{ is a projection on } \underline{H}^{\wedge}. \end{aligned}$$

For each $t > 0$ we also define p^2 operators on $\underset{\sim}{H}^{\wedge}$ $F^{\wedge}(t,\ell,m)$ $\ell,m = 1,\ldots,p$ by

$$F^{\wedge}(t,\ell,m)u^{\wedge} = \chi(t,\ell)E^{\wedge}(t)\chi^{*}(t,m)u^{\wedge} .$$

We observe that each of these operators is bounded with norm equal to one and if $\ell = m$ the operator is in fact a projection. Let us examine in detail the action of $F^{\wedge}(t,\ell,m)$. Let $u^{\wedge} \in \underset{\sim}{H}^{\wedge}$ then

$$\chi^{*}(t,m)u^{\wedge}\cdot(\underline{\theta}) = \begin{cases} t^{n/2}u^{\wedge}(t(\underline{\theta}-\underline{\theta}_m)) & \underline{\theta} \in R_m \\ 0 & \underline{\theta} \notin R_m \end{cases} .$$

Let $\underline{k} \in \underline{Z}_n$

$$(2\pi)^{-n} \int_{\underline{T}_n} \chi^{*}(t,m)u^{\wedge}(\underline{\theta})e^{-i\theta k}d\underline{\theta} = (2\pi)^{-n} \int_{R_m} t^{n/2}u^{\wedge}(t(\underline{\theta}-\underline{\theta}_m))e^{-i\underline{k}\cdot\underline{\theta}}d\underline{\theta} .$$

Making the change of variables $t(\underline{\theta}-\underline{\theta}_m) = \underline{z} = (z_1,\ldots,z_n)$ we obtain

$$(2\pi)^{-n} t^{-n/2} \int_{R_m^t} u\hat{}(\underline{z}) e^{-it^{-1}\underline{k}\cdot\underline{z}} e^{-i\underline{k}\cdot\underline{\theta}_m} d\underline{z} \quad .$$

Therefore

$$E\hat{}(t)(\chi*(t,m)u\hat{})\cdot(\underline{\theta}) = t^{-n/2} \sum_{\underline{k}\in\Omega_t} e^{i\underline{k}\cdot\underline{\theta}} e^{-\underline{k}\cdot\underline{\theta}_m} (2\pi)^{-n} \int_{R_m^t} u\hat{}(\underline{z}) e^{-it^{-1}\underline{k}\cdot\underline{z}} d\underline{z} \quad .$$

Finally applying $\chi(t,\ell)$ we obtain

$$F\hat{}(t,\ell,m)u\hat{}(\underline{x}) = \begin{cases} t^{-n} \sum_{\underline{k}\in\Omega_t} e^{it^{-1}\underline{k}\cdot\underline{x}} e^{i\underline{k}\cdot(\underline{\theta}_\ell-\underline{\theta}_m)} (2\pi)^{-n} \int_{R_m^t} u\hat{}(\underline{z}) e^{-it^{-1}\underline{k}\cdot\underline{z}} d\underline{z} \ , & \underline{x} \in R_\ell^t \\ 0 & , \ \underline{x} \notin R_\ell^t \end{cases} \quad .$$

We note that for $F\hat{}(t,\ell,\ell)$ the exponential factor $e^{i(\underline{\theta}_\ell-\underline{\theta}_m)\cdot\underline{k}} = 1$ while, since the $\underline{\theta}_m$'s are distinct, this is not the case for $F\hat{}(t,\ell,m)$, $\ell \neq m$. An easy computation shows that if $\underline{u}\hat{} = [u_1\hat{},\ldots,u_p\hat{}] \in \underset{\sim}{H}\hat{}$ then

$$\underline{F}\hat{}(t)\underline{u}\hat{} = \left[\sum_{m=1}^{p} F\hat{}(t,1,m)u_m\hat{},\ldots,\sum_{m=1}^{p} F\hat{}(t,p,m)u_m\hat{} \right] \quad .$$

b) On $\underset{\sim}{E}\hat{}$ we define $T\hat{}$ by

$$T\hat{}u\hat{}(\underline{\theta}) = f(\underline{\theta})u\hat{}(\underline{\theta})$$

T is defined on $\underset{\sim}{E}$ by $T = \varphi^{-1}T\hat{}\varphi$. If $u(\underline{k}) \in \underset{\sim}{E}$ then clearly

Setting

$$f(\underline{k}) = (2\pi)^{-n} \int_{\underline{T}} f(\theta) e^{-ik\theta} d\theta$$

we have

$$Tu(\underline{j}) = \sum_{\underline{k} \in \underline{Z}_n} u(\underline{k}) f(\underline{j}-\underline{k}) \quad ,$$

where $f(\underline{k})$, $\underline{k} \in \underline{Z}_n$ is defined in section 2 of I. Next we define $\underline{T}^\wedge(t)$ on $\underset{\approx}{H}^\wedge$ by $\underline{T}^\wedge(t) = \underline{\chi}(t) T^\wedge \underline{\chi}^*(t)$. An easy calculation gives the action of $\underline{T}^\wedge(t)$ on an element $\underline{u}^\wedge = [u_1^\wedge, \ldots, u_p^\wedge] \in \underset{\approx}{H}^\wedge$;

$$\underline{T}^\wedge(t)\underline{u}^\wedge = [f_1^t u_1^\wedge, \ldots, f_p^t u_p^\wedge]$$

where

$$f_m^t(\underline{x}) = \begin{cases} f(t^{-1}\underline{x} + \underline{\theta}_m) & \underline{x} \in R_m^t \\ \\ 0 & \underline{x} \notin R_m^t \end{cases} \quad .$$

We define $S^\wedge(t,m)$, $m = 1,\ldots,p$ on $\underset{\approx}{H}^\wedge$ by

$$S^\wedge(t,m)u^\wedge = t^{\omega} L(t^{-1})^{-1} f_m^t u^\wedge = g_m^t u^\wedge$$

where this equation defines g_m^t . We then define $\underline{S}^\wedge(t)$ on $\underset{\approx}{H}^\wedge$ by

$$S^\wedge(t)\underline{u}^\wedge = [S^\wedge(t,1)u_1^\wedge, \ldots, S^\wedge(t,p)u_p^\wedge] \ .$$

Note that $\underline{S}^\wedge(t) = t^{\omega} L(t^{-1})^{-1} \underline{T}^\wedge(t)$ and that for every t $S^\wedge(t)$ is a bounded operator on $\underset{\approx}{H}^\wedge$.

c) $\underline{S}^\wedge$ is defined on $\underset{\approx}{H}^\wedge$ as follows. Let

$$g_m(\underline{x}) = |\underline{x}|^{\omega} \phi_m(\underline{x}) \ , \ m = 1,\ldots,p \quad .$$

Define $S^\wedge(m): \underset{\sim}{H}^\wedge \to \underset{\sim}{H}^\wedge$ by

$$S^\wedge(m)u^\wedge = g_m u^\wedge \ , \qquad m = 1,\ldots,p \ .$$

Finally we set

$$\underline{S}^\wedge \underline{u}^\wedge = [S^\wedge(1)u_1^\wedge,\ldots,S^\wedge(p)u_p^\wedge] \quad .$$

$\underline{S}^\wedge$ is a positive self-adjoint operator defined for all $\underline{u}^\wedge \in \underset{\approx}{H}^\wedge$ for which

$$\int_{\underline{R}_n} |g_m(\underline{x})u_m^\wedge(\underline{x})|^2 d\underline{x} < \infty \ , \quad m = 1,\ldots,p \quad .$$

d) We define a projection on $\underset{\approx}{H}^\wedge$ as follows. Let F be the projection defined in $\underset{\sim}{H}$ by

$$Fu\cdot(\underline{y}) = \begin{cases} u(\underline{y}), & \underline{y} \in \Omega \\ 0, & \underline{y} \notin \Omega \end{cases}$$

and define $\underline{F}$ on $\underset{\approx}{H}$ by

$$\underline{F}\ \underline{u} = [Fu_1,\ldots,Fu_p] \quad .$$

$\underline{F}^\wedge$ is defined on $\underset{\approx}{H}^\wedge$ by $\underline{F}^\wedge = \Psi F \Psi^{-1}$. An easy computation gives

$$\underline{F}^\wedge\underline{u}^\wedge(\underline{x}) = \left[\int_\Omega (\Psi^{-1}u_1^\wedge)(\underline{y})e^{i\underline{x}\cdot\underline{y}}\, d\underline{y},\ldots,\int_\Omega(\Psi^{-1}u_p^\wedge)(\underline{y})e^{i\underline{x}\cdot\underline{y}} d\underline{y}\right] .$$

Let $0 < \lambda(t.1) \leq \lambda(t,2) \leq \cdots$ be the eigen values of the operators

$$E(t)TE(t)|_{E(t)\underset{\sim}{E}} \quad ,$$

$$E^\wedge(t)T^\wedge E^\wedge(t)|_{E^\wedge(t)\underset{\sim}{E}^\wedge} \quad ,$$

$$\underline{F}^\wedge(t)\underline{T}^\wedge(t)\underline{F}^\wedge(t)|_{\underline{F}^\wedge(t)\underset{\approx}{H}^\wedge} \ ,$$

where these symbols are to be read "$E(t)TE(t)$ restricted to $E(t)\underset{\sim}{E}$," etc. The eigen values of

$$\underline{F}^{\wedge}(t)\underline{S}^{\wedge}(t)\underline{F}^{\wedge}(t)|_{\underline{F}^{\wedge}(t)\underset{\approx}{H}^{\wedge}}$$

in increasing order are then

$$t^{\omega}L(t^{-1})^{-1}\lambda(t,1) \leq t^{\omega}L(t^{-1})^{-1}\lambda(t,2) \leq \cdots \quad .$$

In the Sections 3-7 we show that the operators $\underline{F}^{\wedge}, \underline{S}^{\wedge}\underline{F}^{\wedge}(t)$ and $\underline{S}^{\wedge}$ satisfy the assumptions of our perturbation theory. As in III this will yield our desired result.

3. Convergence of $\underline{S}^{\wedge}(t)^{\frac{1}{2}}$ to $(\underline{S}^{\wedge})^{\frac{1}{2}}$

The arguments here are so much like those given in §3 of III that we merely record the final result.

Theorem 3a. $(\underline{S}^{\wedge})^{\frac{1}{2}}$ is the closure of the strong limit of $(\underline{S}^{\wedge}(t))^{\frac{1}{2}}$ as $t \to \infty$.

4. Convergence of $\underline{F}^{\wedge}(t)$ to $\underline{F}^{\wedge}$

Throughout the present section we suppose that Ω is any set of finite positive measure for which $\partial\Omega$ has measure 0. This degree of generality is needed in a later chapter.

Let $\underset{\sim}{D}(\Omega)$ be the set of functions in $\underset{\sim}{H}$ which are infinitely differentiable with compact support continaed in Ω. The set $\underset{\approx}{D}^{\wedge}(\Omega) = \Phi\underset{\sim}{D}(\Omega)$. $\underset{\approx}{D}(\Omega)(\underset{\approx}{D}^{\wedge}(\Omega))$ is defined as the collection of elements of $\underset{\approx}{H}(\underset{\approx}{H}^{\wedge})$ with each component an element of $\underset{\sim}{D}(\Omega)\underset{\approx}{D}^{\wedge}(\Omega))$. We also define corresponding sets for Ω', the complement of Ω, and denote them by $\underset{\sim}{D}(\Omega')$, $\underset{\sim}{D}(\Omega')$, etc. Since $\partial\Omega$ is assumed to have measure zero the vector space $\underset{\sim}{D}(\Omega) + \underset{\sim}{D}(\Omega')(\underset{\sim}{D}^{\wedge}(\Omega') + \underset{\sim}{D}^{\wedge}(\Omega'))$ is dense in $\underset{\sim}{H}(\underset{\sim}{H}^{\wedge})$.

Lemma 4a. If $u^{\wedge}(\underline{x}) \in \underset{\sim}{H}^{\wedge}$ and $u^{\wedge}(\underline{x}) = O(|\underline{x}|^{-s})$ for every positive integer s then we have:

i.

(1) $$\sum_{\underline{j} \neq \underline{0}} \{(2\pi)^{-n} \int_{R_{\ell}^{t}} |u^{\wedge}(\underline{x} + \underline{j}2\pi t)|^{2} d\underline{x}\}^{\frac{1}{2}} = O(t^{-r})$$

for every positive integer r ;

ii. if $\ell \neq m$

(2) $$\sum_{\underline{j} \in \underline{Z}_n} [(2\pi)^{-n} \int_{R_{\ell}^{t}} |u^{\wedge}(\underline{x} + 2\pi t\underline{j} + t(\underline{\theta}_{\ell} - \underline{\theta}_{m}))|^{2} d\underline{x}]^{\frac{1}{2}} = O(t^{-r})$$

for every positive integer r.

Proof. i. We clearly have for all $j \neq 0$

$$\mathrm{dist}(R_{\ell}^{1} + 2\pi\underline{j}, \underline{0}) \geq q|\underline{j}| \text{ for some } q > 0.$$

Therefore

$$\int_{R_{\ell}^{t}} |u^{\wedge}(\underline{x} + 2\pi t\underline{j})|^{2} dx = \int_{R_{\ell}^{t}+2\pi t\underline{j}} |u^{\wedge}(\underline{y})|^{2} d\underline{y} \leq \int_{|\underline{y}| > qt|\underline{j}|} |u^{\wedge}(\underline{y})|^{2} d\underline{y} .$$

Since $u^{\wedge}(\underline{x}) = O(|\underline{x}|^{-s})$ as $|\underline{x}| \to \infty$ for every positive integer s we have

$$\{\int_{R_{\ell}^{t}} |u^{\wedge}(\underline{x} + 2\pi t\underline{j})|^{2} d\underline{x}\}^{\frac{1}{2}} = O(t^{-r})|\underline{j}|^{-r}$$

for every $r > 0$. This implies

$$\sum_{\underline{j} \neq \underline{0}} [(2\pi)^{-n} \int_{R^t_\ell} |u\hat{}(\underline{x} + 2\pi t\underline{j})|^2 dx]^{\frac{1}{2}} \leq O(t^{-r}) \sum_{\underline{j} \neq \underline{0}} |\underline{j}|^{-r}$$

for every $r > 0$ and (1) holds.

ii. It is clear that (2) will follow from an analogous argument if we can show that for all $\underline{j} \in \underline{Z}_n$

(3)
$$\mathrm{dist}(R^1_\ell + 2\pi\underline{j} + (\underline{\theta}_\ell - \underline{\theta}_m), \underline{0}) > q(|\underline{j}| + 1)$$

for some $q > 0$. We first show that $\underline{\theta}_m - \underline{\theta}_\ell \notin R^1_\ell$.

Indeed if $R_\ell = [\underline{a}_\ell, \underline{b}_\ell]$ where $\underline{a}_\ell = (a_{\ell_1}, \ldots, a_{\ell_n})$, $\underline{b}_\ell = (b_{\ell_1}, \ldots, b_{\ell_n})$ then

$$R^1_\ell = [\underline{a}_\ell - \underline{\theta}_\ell, \underline{b}_\ell - \underline{\theta}_\ell] \quad .$$

If $\underline{\theta}_m - \underline{\theta}_\ell \in R^1_\ell$ then for $1 \leq j \leq n$

$$a_{\ell_j} - \theta_{\ell_j} \leq \theta_{m_j} - \theta_{\ell_j} \leq b_{\ell_j} - \theta_{\ell_j}$$

or equivalently

$$a_{\ell_j} \leq \theta_{m_j} \leq b_{\ell_j} \quad .$$

This implies that $\underline{\theta}_m \in R_\ell$ which is a contradiction.

Thus we conclude that

$$\underline{\theta}_m - \underline{\theta}_\ell = \sigma(1,\ell)(\underline{\theta}_m) \in \sigma(1,\ell)\underline{T}_n \backslash R^1_\ell \quad .$$

In view of this we have the following configuration in the case n = 2:

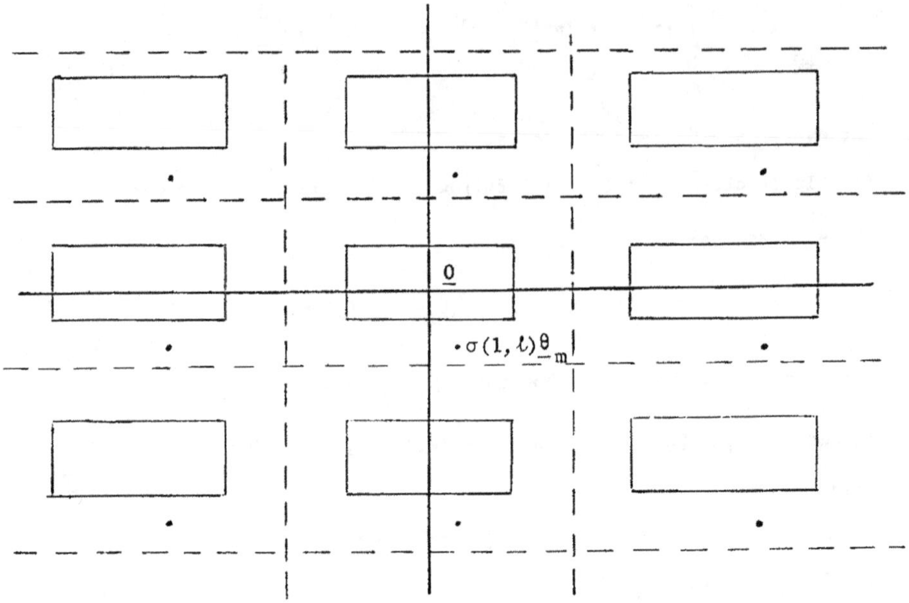

Here each rectangle bounded with a solid line represents $R_\ell^1 + 2\pi\underline{j}$ for some $\underline{j}$, while each rectangle bounded by a broken line represents $\sigma(1,\ell)\underline{T}_2 + 2\pi\underline{j}$ for some $\underline{j}$. We have

$$\mathrm{dist}(R_\ell^1 + \underline{\theta}_\ell - \underline{\theta}_m + 2\pi\underline{j}, \underline{0}) = \mathrm{dist}(\sigma(1,\ell)\underline{\theta}_m, R_\ell^1 + 2\pi\underline{j}) \quad .$$

From this we see that (3) holds.

Theorem 4b. Let s be any positive integer. We have:

i. if $u^\wedge \in \underset{\sim}{D}^\wedge(\Omega)$ then for $\ell = 1,\ldots,p$

$$\|F^\wedge(t,\ell,\ell)u^\wedge - u^\wedge\|_{\underset{\sim}{H}^\wedge} = O(t^{-s}) \qquad \text{as} \quad t \to \infty \quad ,$$

$$\|F^\wedge(t,\ell,\ell)u^\wedge - u^\wedge\|_\infty = O(t^{-s}) \qquad \text{as} \quad t \to \infty \quad ;$$

ii. if $u^\wedge \in \underset{\sim}{D}^\wedge(\Omega')$ then for $\ell = 1,\ldots,p$

$$\|F^\wedge(t,\ell,\ell)u^\wedge\|_{\underset{\sim}{H}^\wedge} = O(t^{-s}) \quad \text{as} \quad t \to \infty ,$$

$$\|F^\wedge(t,\ell,\ell)u^\wedge\|_\infty = O(t^{-s}) \quad \text{as} \quad t \to \infty ;$$

iii. if $u^\wedge \in \underset{\sim}{D}^\wedge(\Omega)$ then for $1 \leq \ell,\ m \leq p,\ \ell \neq m$

$$\|F^\wedge(t,\ell,m)u^\wedge\|_{\underset{\sim}{H}^\wedge} = O(t^{-s}) \quad \text{as} \quad t \to \infty ,$$

$$\|F^\wedge(t,\ell,m)u^\wedge\|_\infty = O(t^{-s}) \quad \text{as} \quad t \to \infty ;$$

iv. if $u^\wedge \in \underset{\sim}{D}^\wedge(\Omega')$ then for $1 \leq \ell,\ m \leq p,\ \ell \neq m$

$$\|F^\wedge(t,\ell,m)u^\wedge\|_{\underset{\sim}{H}^\wedge} = O(t^{-s}) \quad \text{as} \quad t \to \infty ,$$

$$\|F^\wedge(t,\ell,m)u^\wedge\|_\infty = O(t^{-s}) \quad \text{as} \quad t \to \infty .$$

Proof i. Recall

$$F^\wedge(t,\ell,\ell)u^\wedge\cdot(\underline{x}) = \begin{cases} t^{-n} \sum\limits_{\underline{k}\in\Omega_t} e^{it^{-1}\underline{k}\cdot\underline{x}}(2\pi)^{-n} \int\limits_{R_\ell^t} u^\wedge(\underline{z})e^{it^{-1}\underline{k}\cdot\underline{z}}d\underline{z} , & \underline{x} \in R_\ell^t \\ 0 & , \ \underline{x} \notin R_\ell^t \end{cases} .$$

Let $u^\wedge \in \underset{\sim}{D}^\wedge(\Omega)$; then for $\underline{x} \in R_\ell^t$

$$F^\wedge(t,\ell,\ell)u^\wedge(\underline{x}) = t^{-n} \sum_{\underline{k}\in\Omega_t} e^{it^{-1}\underline{k}\cdot\underline{x}}(2\pi)^{-n} \int_{\underline{R}_n} u^\wedge(\underline{z})e^{-it^{-1}\underline{k}\cdot\underline{z}}\, d\underline{z}$$

$$- t^{-n} \sum_{\underline{k}\in\Omega_t} e^{it^{-1}\underline{k}\cdot\underline{x}}(2\pi)^{-n} \int_{\underline{R}_n \setminus R_\ell^t} u^\wedge(\underline{z})e^{-t^{-1}\underline{k}\cdot\underline{z}}d\underline{z} ,$$

$$= u_t^\wedge(x) - v_t^\wedge(\underline{x})$$

where $u_t^{\wedge}(\underline{x})$ and $v_t^{\wedge}(\underline{x})$ are defined by the right hand side of the above equation for $\underline{x} \in R_\ell^t$ and as 0 for $\underline{x} \notin R_\ell^t$. For $\underline{x} \in R_\ell^t$ we have

$$u_t^{\wedge}(\underline{x}) = \sum_{\underline{k} \in \Omega_t} e^{-it^{-1}\underline{k}\cdot\underline{x}} t^{-n} u(t^{-1}\underline{k}), \qquad u = \psi^{-1}u^{\wedge} ,$$

(3)

$$= \sum_{\underline{k} \in \underline{Z}_n} e^{-it^{-1}\underline{k}\cdot\underline{x}} t^{-n} u(t^{-1}\underline{k}) ,$$

since $u(t^{-1}\underline{k}) = 0$ if $\underline{k} \notin \Omega_t$. Consider the function

$$w_t^{\wedge}(\underline{x}) = \sum_{\underline{j} \in Z_n} u^{\wedge}(\underline{x} + 2\pi\underline{j}t) \qquad \underline{x} \in \underline{R}_n .$$

$w_t^{\wedge}(\underline{x})$ is periodic with period $2\pi t$. For $\underline{k} \in \underline{Z}_n$ we have

$$(2\pi)t^{-n} \int_{\sigma(t,\ell)\underline{T}_n} w_t^{\wedge}(\underline{x}) e^{-it^{-1}\underline{k}\cdot\underline{x}} d\underline{x} =$$

$$= (2\pi t)^{-n} \sum_{\underline{j} \in \underline{Z}_n} \int_{\sigma(t,\ell)\underline{T}_n} u^{\wedge}(\underline{x} + 2\pi\underline{j}t) e^{-it^{-1}\underline{k}\cdot\underline{x}} d\underline{x}$$

$$= (2\pi)^{-n} t^{-n} \int_{\underline{R}_n} u^{\wedge}(\underline{z}) e^{-it^{-1}\underline{k}\cdot\underline{z}} d\underline{z} = t^{-n} u(t^{-1}\underline{k}) ,$$

which implies that for $\underline{x} \in \underline{R}_n$

$$w_t^{\wedge}(\underline{x}) = \sum_{\underline{k}\in \underline{Z}_n} e^{it^{-1}\underline{k}\cdot\underline{x}} t^{-n} u(t^{-1}\underline{k}) \quad .$$

Comparing this with (3) we see that

$$u_t^{\wedge}(\underline{x}) = \sum_{\underline{j}\in \underline{Z}_n} u^{\wedge}(\underline{x}+2\pi\underline{j}t)$$

(4)

$$= u^{\wedge}(\underline{x}) + \sum_{\underline{j}\neq 0} u^{\wedge}(\underline{x}+2\pi\underline{j}t) \qquad \text{for} \quad \underline{x} \in R_{\ell}^t \quad .$$

Note: This relation does not hold outside R_{ℓ}^t since $u_t^{\wedge}(\underline{x})$ is 0 for $\underline{x} \in \underline{R}_n \backslash R_{\ell}^t$. This implies that

$$\|u_t^{\wedge} - u^{\wedge}\|_{\underset{\sim}{H}^{\wedge}} \leq \|u^{\wedge}|_{\underline{R}_n\backslash R_{\ell}^t}\|_{\underset{\sim}{H}^{\wedge}} + \sum_{\underline{j}\neq \underline{0}} \|u^{\wedge}(\underline{x}+2\pi\underline{j}t)|_{R_{\ell}^t}\|_{\underset{\sim}{H}^{\wedge}} \quad .$$

It follows from Lemma 4a that as $t \to \infty$

$$\sum_{\underline{j}\neq \underline{0}} \|u^{\wedge}(\underline{x}+2\pi\underline{j}t)|_{R_{\ell}^t}\|_{\underset{\sim}{H}^{\wedge}} = 0(t^{-s})$$

for every positive s. Since $|u^{\wedge}(\underline{x})| = 0(|\underline{x}|^{-r})$ as $\underline{x} \to \infty$ for every positive r we have

$$\|u^{\wedge}|_{\underline{R}_n\backslash R_{\ell}^t}\|_{\underset{\sim}{H}^{\wedge}} = 0(t^{-s})$$

for every positive s. Combining these results we have

$$\|u^{\wedge} - u_t^{\wedge}\|_{\underset{\sim}{H}^{\wedge}} = 0(t^{-s}) \qquad \text{as} \quad t \to \infty$$

for every positive s, etc.

It is apparent that

$$(2\pi)^{-n} \int_{\underline{R}_n \setminus R_\ell^t} u^\wedge(\underline{z})^{-it^{-1}\underline{k}\cdot\underline{z}} d\underline{z} = O(t^{-r})$$

for every r. Since the number of lattice points in Ω_t is $O(t^n)$ -- it is at this point that the assumption that Ω is bounded enters -- we see that

$$|v_t(\underline{x})| = O(t^{-r}) \quad \text{as} \quad t \to \infty$$

uniformly in $\underline{x}$. Since $v_t(\underline{x})$ is zero outside R_ℓ^t it follows that

$$\|v_t(\underline{x})\|_{R_\ell^t}\|_{\underset{\sim}{H}^\wedge} = O(t^{-s})$$

and our first assertion follows. A very slight variation of these arguments, see §4 of III, proves our second assertion

ii. Let $u^\wedge \in \underset{\sim}{D}^\wedge(\Omega')$. For $\underline{x} \in R_\ell^t$ we have (since $u(t^{-1}\underline{k}) = 0$ if $\underline{k} \in \Omega_t$)

$$F^\wedge(t,\ell,\ell)u^\wedge(\underline{x}) = -t^{-n} \sum_{\underline{k}\in\Omega_t} e^{it^{-1}\underline{k}\cdot\underline{x}} (2\pi)^{-n} \int_{\underline{R}_n \setminus R_\ell^t} u^\wedge(\underline{z}) e^{-it^{-1}\underline{k}\cdot\underline{z}} d\underline{z} \quad .$$

For $\underline{x} \in \underline{R}_n \setminus R_\ell^t$, $F^\wedge(t,\ell,\ell)u^\wedge(\underline{x}) = 0$. The argument used above shows that in this case

$$\|F^\wedge(t,\ell,\ell)u^\wedge\|_{\underset{\sim}{H}^\wedge} = O(t^{-s})$$

for every positive s, etc.

iii. We recall that

$$F^{\wedge}(t,\ell,m)u^{\wedge}(\underline{x}) = t^{-n}\sum_{\underline{k}\in\Omega_t} e^{it^{-1}\underline{k}\cdot\underline{x}} e^{i\underline{k}\cdot(\underline{\theta}_\ell-\underline{\theta}_m)}(2\pi)^{-n}\int_{R_m^t} u^{\wedge}(\underline{z})e^{-it^{-1}\underline{k}\cdot\underline{z}}d\underline{z}$$

for $\underline{x} \in R_\ell^t$; $F^{\wedge}(t,\ell,m)u^{\wedge}(\underline{x}) = 0$ for $\underline{x} \in \underline{R}_n\setminus R_\ell^t$. Let $u^{\wedge} \in \underset{\sim}{\underline{D}}^{\wedge}(\Omega)$. Then for $\underline{x} \in R_\ell^t$

$$F^{\wedge}(t,\ell,m)u^{\wedge}(\underline{x}) = \sum_{\underline{k}\in\underline{Z}_n} e^{it^{-1}\underline{k}\cdot\underline{x}} e^{i\underline{k}\cdot(\underline{\theta}_\ell-\underline{\theta}_m)}(2\pi t)^{-n}\int_{\underline{R}_n} u^{\wedge}(\underline{z})e^{-it^{-1}\underline{k}\cdot\underline{z}}d\underline{z}$$

$$- t^{-n}\sum_{\underline{k}\in\Omega_t} e^{it^{-1}\underline{k}\cdot\underline{x}} e^{i\underline{k}\cdot(\underline{\theta}_\ell-\underline{\theta}_m)}(2\pi)^{-n}\int_{\underline{R}_n\setminus R_m^t} u^{\wedge}(\underline{z})e^{-it^{-1}\underline{k}\cdot\underline{z}}d\underline{z}$$

$$= v_t^{\wedge}(\underline{x}+t(\underline{\theta}_\ell-\underline{\theta}_m)) - v_t^{\wedge}(\underline{x}) \quad ,$$

where $u_t^{\wedge}$ is as before and where $v_t^{\wedge}(\underline{x})$ is defined for $\underline{x} \in \underline{R}_n$ by the above relation. Using (4) we see that

$$\|u_t^{\wedge}(\underline{x}+t(\underline{\theta}_\ell-\underline{\theta}_m))|_{R_\ell^t}\|_{\underset{\sim}{H}^{\wedge}} \le \sum_{\underline{j}\in\underline{Z}_n}' \|u^{\wedge}(\underline{x}+2\pi\underline{j}t+t(\underline{\theta}_\ell-\underline{\theta}_m))|_{R_\ell^t}\|_{\underset{\sim}{H}^{\wedge}} \ .$$

By ii. of Lemma 4a

$$\|u_t^{\wedge}(\underline{x}+t(\underline{\theta}_\ell-\underline{\theta}_m))|_{R_\ell^t}\|_{\underset{\sim}{H}^{\wedge}} = O(t^{-s})$$

for every positive s. An argument like the one given earlier shows that

$$\|v_t^{\wedge}(\underline{x})|_{R_\ell^t}\|_{\underset{\sim}{H}^{\wedge}} = O(t^{-s})$$

for every positive s, etc.

iv. Let $u^{\wedge} \in \underset{\sim}{D}^{\wedge}(\Omega')$. Then for $\underline{x} \in R^{t}_{\ell}$

$$F^{\wedge}(t,\ell,m)u^{\wedge}(\underline{x}) = -t^{-n} \sum_{\underline{k} \in \Omega_t} e^{it^{-1}\underline{k}\cdot\underline{x}} e^{i\underline{k}\cdot(\underline{\theta}_{\ell}-\underline{\theta}_m)} (2\pi)^{-n} \int_{\underline{R}_n \setminus R^t_m} u^{\wedge}(\underline{z}) e^{-it^{-1}\underline{k}\cdot\underline{z}} d\underline{z} \ .$$

By the argument used in iii we clearly have

$$\|F^{\wedge}(t,\ell,m)u^{\wedge}\|_{\underset{\sim}{H}^{\wedge}} = O(t^{-s}) \quad ,$$

$$\|F^{\wedge}(t,\ell,m)u^{\wedge}\|_{\infty} = O(t^{-s}) \quad ,$$

for every positive integer s.

Lemma 4c. $\underline{F}^{\wedge}$ is the strong limit of $\underline{F}^{\wedge}(t)$ as $t \to +\infty$.

Proof. If $\underline{u}^{\wedge} \in \underset{\approx}{D}^{\wedge}(\Omega) + \underset{\approx}{D}^{\wedge}(\Omega')$ then the above results imply that

$$\|\underline{F}^{\wedge}(t)\underline{u}^{\wedge} - \underline{F}^{\wedge}\underline{u}^{\wedge}\|_{\underset{\approx}{H}^{\wedge}} \to 0 \quad \text{as } t \to +\infty \ .$$

But $\underset{\approx}{D}^{\wedge}(\Omega) + \underset{\approx}{D}^{\wedge}(\Omega')$ is a dense subset of $\underset{\approx}{H}^{\wedge}$, etc.

5. Convergence of $\underline{S}^{\wedge}(t)^{\frac{1}{2}}\underline{F}^{\wedge}(t)$ to $(\underline{S}^{\wedge})^{\frac{1}{2}}\underline{F}^{\wedge}$, Part I

The proofs of the following results follow so closely the pattern established in §5 of III that there is no loss in omitting them.

Lemma 5a. If $u^{\wedge} \in \underset{\sim}{D}^{\wedge}(\Omega)$ then for $1 \leq \ell \leq p$

$$\|S^{\wedge}(t,\ell)^{\frac{1}{2}}F^{\wedge}(t,\ell,\ell)u^{\wedge} - S^{\wedge}(\ell)^{\frac{1}{2}}F^{\wedge}u^{\wedge}\|_{\underset{\sim}{H}^{\wedge}} \to 0$$

as $t \to \infty$.

Lemma 5b. If $u^{\wedge} \in \underset{\sim}{D}^{\wedge}(\Omega)$ then for $1 \leq \ell, m \leq p \quad \ell \neq m$

$$\|S^{\wedge}(t,\ell)^{\frac{1}{2}}F^{\wedge}(t,\ell,m)u^{\wedge}\|_{\underset{\sim}{H}^{\wedge}} \to 0 \qquad \text{as } t \to \infty \quad .$$

Lemma 5c. If $\underline{u}^{\wedge} \in \underset{\approx}{D}^{\wedge}(\Omega)$ then

$$\|\underline{S}^{\wedge}(t)^{\frac{1}{2}}\underline{F}^{\wedge}(t)\underline{u}^{\wedge} - (\underline{S}^{\wedge})^{\frac{1}{2}}\underline{u}^{\wedge}\|_{\underset{\approx}{H}^{\wedge}} \to 0 \quad \text{as } t \to \infty \quad .$$

6. Convergence of $\underline{S}^{\wedge}(t)^{\frac{1}{2}}\underline{F}^{\wedge}(t)$ to $(\underline{S}^{\wedge})^{\frac{1}{2}}$ on $\underset{\approx}{M}^{\wedge}$, Part II

The proofs here follow exactly the pattern established in §6 of III. We therefore again record only the final result.

Theorem 6a. $(\underline{S}^{\wedge})^{\frac{1}{2}}\underline{F}^{\wedge}$ is the closure of $\underline{S}^{\wedge}(t)^{\frac{1}{2}}\underline{F}^{\wedge}(t)$ as $t \to \infty$ on $\underset{\approx}{M}^{\wedge}$.

7. The Asymptotic Formula, I

Let $\underline{S}_{\underline{F}}^{\wedge}$ be constructed from $\underline{F}^{\wedge}$ and $S^{\wedge}$ as in Section 1 of Chapter II. Let

$$\underline{S}_{\underline{F}}^{\wedge} = \int_{0-}^{\infty} \lambda \, dC(\lambda)$$

be the spectral resolution of $\underline{S}_{\underline{F}}^{\wedge}$ on $\underset{\approx}{M}^{\wedge} = \underline{F}^{\wedge}\underset{\approx}{H}^{\wedge}$, and let

$$\underline{S}_{\underline{F}}^{\wedge}(t) = \int_{0-}^{\infty} \lambda \, dC(t,\lambda)$$

be the spectral resolution of $\underline{S}_{\underline{F}}^{\wedge}(t)$. It follows from Section 1-4 and Corollary 4b of II that

(1) $$C(t,\lambda) \to C(\lambda) \qquad \text{as } t \to \infty \; ,$$

for every $\lambda \notin \sigma_p(\underline{S}_{\underline{F}}^{\wedge})$.

Lemma 7a. For $0 < t < \infty$ let $\{\underline{u}^{\wedge}(t,\underline{x})\}$ be a family of functions such that

i. $\underline{u}^{\wedge}(t,\underline{x}) \in \underline{F}^{\wedge}(t)\underline{H}^{\wedge}$ all $t > 0$,

ii. $\|\underline{u}^{\wedge}(t,\underline{x})\|_{\underline{H}^{\wedge}} = 1$ all $t > 0$.

If $\underline{u}^{\wedge}(t,\underline{x}) = (u^{\wedge}(t,1,x),\ldots,u^{\wedge}(t,p,\underline{x}))$ then for each m, $1 \leq m \leq p$, $\{u^{\wedge}(t,m,\underline{x})\}$ $t > 0$ is, for t large, a uniformly bounded and equicontinuous family of functions.

Proof. Since $\underline{u}^{\wedge}(t,\underline{x}) \in \underline{F}^{\wedge}(t)\underline{H}^{\wedge}$ we have $\underline{u}^{\wedge}(t,\underline{x}) = \underline{F}^{\wedge}(t)\underline{u}^{\wedge}(t,\underline{x})$. Hence

(2) $$u^{\wedge}(t,k,\underline{x}) = \sum_{m=1}^{p} F^{\wedge}(t,k,m)u^{\wedge}(t,m)\cdot(\underline{x}) \qquad k = 1,\ldots,p \ ,$$

where

$$F^{\wedge}(t,k,m)u^{\wedge}(t,m)\cdot(\underline{x}) = \begin{cases} \displaystyle\sum_{\underline{j}\in\Omega_t} e^{it^{-1}\underline{j}\cdot\underline{x}+i\underline{j}(\underline{\theta}_k-\underline{\theta}_m)}u(m,t,\underline{j}) & \underline{x} \in R_k^t \\ 0 & x \notin R_k^t \end{cases}$$

and

$$u(m,t,\underline{j}) = (2\pi t)^{-n}\int_{R_m^t} u^{\wedge}(t,m,\underline{z})e^{-it^{-1}\underline{j}\cdot\underline{z}}\,d\underline{z} \quad .$$

Let k be fixed, we will show that there exist constants M and M' such that for all large t

(3) $$|u^{\wedge}(t,k,\underline{x})| \leq M, \ \underline{x} \in \underline{R}_n$$

and

(4) $$\left|\frac{\partial}{\partial x_r} u^{\wedge}(t,k,\underline{x})\right| \leq M' \ , \ \underline{x} \in \underline{R}_n \ , \quad r = 1,\ldots,n \ .$$

By Bessel's inequality

$$\sum_{\underline{j} \in \Omega_t} |u(m,t,\underline{j})|^2 \leq t^{-n} \| u\hat{}(t,m,\underline{z}) \|^2_{\underset{\sim}{H}\hat{}} = t^{-n}$$

and therefore

$$|F\hat{}(t,k,m)u\hat{}(t,m)\cdot(x)|^2 \leq \sum_{\underline{j} \in \Omega_t} 1 \sum_{\underline{j} \in \Omega_t} |u(m,t,\underline{j})|^2$$

$$\leq t^{-n} \sum_{\underline{j} \in \Omega_t} 1 \quad .$$

Since Ω is bounded the number of points in Ω_t is $O(t^n)$ and we have proved (3).

From (2) we have for $r = 1,\ldots,n$

$$\frac{\partial}{\partial x_r} u\hat{}(t,k,\underline{x}) = \sum_{m=1}^{p} \frac{\partial}{\partial x_r} [F\hat{}(t,k,m)u\hat{}(t,m)\cdot(\underline{x})] \quad .$$

By Schwarz's inequality

$$\left|\frac{\partial}{\partial x_r} F\hat{}(t,k,m)u\hat{}(t,m)\cdot(\underline{x})\right|^2 \leq \sum_{\underline{j} \in \Omega_t} \left|\frac{j_r}{t}\right|^2 \sum_{\underline{j} \in \Omega_t} |u(m,t,\underline{j})|^2 ,$$

$$\leq t^{-n} \sum_{\underline{j} \in \Omega_t} \left|\frac{j_r}{t}\right|^2 \quad ,$$

where j_r is the r-th component of $\underline{j}$. j_r/t is then the r-th coordinate of a point in Ω. Since Ω is bounded and since the number of points in Ω_t is $O(t^n)$ we have established (4).

Lemma 7b. Let $\{\underline{u}\hat{}(t,\underline{x})\}$ $t > 0$ be a family of functions such that for all $t > 0$,

i. $\underline{u}^{\wedge}(t,\underline{x}) \in \underline{F}^{\wedge}(t)\underset{\approx}{H}^{\wedge}$,

ii. $\|u^{\wedge}(t,\underline{x})\|_{\underset{\approx}{H}^{\wedge}} = 1$,

iii. $\langle \underline{S}^{\wedge}(t,F)\underline{u}^{\wedge}(t) | \underline{u}^{\wedge}(t)\rangle_{\underset{\approx}{H}^{\wedge}} \leq M < \infty$.

We assert that if $\underline{u}^{\wedge}(t) \rightharpoonup \underline{u}^{\wedge}$ as $t \to \infty$ in $\underset{\sim}{P}_1$, a subsequence of $\underset{\sim}{P}$, then $\underline{u}^{\wedge} \neq 0$.

Proof. The proof of this lemma is virtually that of Lemma 7b of III.

Since the operator $\underline{S}_{\underline{F}^{\wedge}}^{\wedge}$ is exactly the same as in §7 of III, we may define $\{\mu(k)\}_1^{\infty}$ in the same way. Applying the pertubation theory of Chapter II, the assumptions of which we have verified we obtain the following.

Theorem 7c. Under assumptions i.-v. of §1 we have

$$\lim_{t \to \infty} t^{\omega} L(t^{-1})^{-1} \lambda(k,t) = \mu(k) \qquad k = 1,2,\ldots$$

where, for t sufficiently large, $\lambda(k,t)$ is the k-th eigen value of $E^{\wedge}(t)T^{\wedge}E^{\wedge}(t)$ restricted to $E^{\wedge}(t)\underset{\sim}{H}^{\wedge}$ counting from the bottom up. We observe here the convention that eigen values are repeated according to their multiplicities.

8. The Asymptotic Formula, II

We can make the same extension here that we made at the end of III; that is, we may suppose that $f(\underline{\theta})$ has, in addition to q zeros of the type we have previously considered, p-q zeros of lower order. This all goes exactly as before and we omit both the statement and the demonstration.

Chapter V

HILBERT SPACE THEORY - LARGE EIGEN VALUES

1. A Perturbation Theorem

This chapter is devoted to the Hilbert space results which are needed in Chapter VI to develop in general form the theory sketched in §2 of I.

Let $\underset{\sim}{H}$ be a separable Hilbert space on which are defined the following families of operators.

i. $F(t)$, $0 < t < \infty$, and F are bounded linear operators on $\underset{\sim}{H}$ satisfying

$$F(t)^* \to F^* \quad t \to \infty \quad .$$

Note that as a consequence of the uniform boundedness principle there exists a constant M such that

$$\|F^*(t)\| \leq M, \ \|F(t)\| \leq M$$

for all sufficiently large t.

ii. $U(t)$, $0 < t < \infty$, and U are not necessarily bounded self-adjoint operators such that $\underset{\sim}{R}[F(t)] \subset \underset{\sim}{D}[U(t)]$, $\underset{\sim}{R}(F) \subset \underset{\sim}{D}[U]$, which implies that the operators $U(t)F(t)$ and UF are bounded. We assume that

$$U(t)F(t) \to UF \quad \text{as} \quad t \to \infty \ .$$

The uniform boundedness principle implies that there exists a constant M such that

$$\|U(t)F(t)\| \leq M$$

for all sufficiently large t.

iii. $V(t)$, $0 < t < \infty$, and V are not necessarily bounded self-adjoint operators such that $\underset{\sim}{R}[F(t)] \subset \underset{\sim}{D}[V(t)]$, $\underset{\sim}{R}[F] \subset \underset{\sim}{D}[V]$ and

$$V(t)F(t) \rightarrow VF \qquad \text{as} \quad t \rightarrow \infty \quad .$$

Again, for M suitably chosen,

$$\|V(t)F(t)\| \leq M$$

for t sufficiently large.

iv. We assume that

$$\langle U(t)F(t)u | V(t)F(t)v\rangle = \langle V(t)F(t)u | U(t)F(t)v\rangle \tag{1}$$

for every $u,v \in \underset{\sim}{H}$, and all t, $0 < t < \infty$.

v. We assume that U is the closure of the strong limit of $U(t)$ as $t \rightarrow \infty$.

Alternatively we could assume that V is the closure of the strong limit of $V(t)$ as $t \rightarrow \infty$.

It follows from assumptions ii and iii that the mapping of $\underset{\sim}{H} \times \underset{\sim}{H}$ into $\underline{C}$, the complex numbers, which sends $\{u,v\}$ into

$$\langle U(t)F(t)u | V(t)F(t)v\rangle$$

is sesquilinear and bounded. Therefore, by a familiar elementary result, there exists a bounded operator which we denote by $S_F(t)$ such that

$$\langle S_F(t)u | v\rangle = \langle U(t)F(t)u | V(t)F(t)v\rangle \quad .$$

Using assumption iv. we see that

$$\langle S_F(t)u|v\rangle = \langle V(t)F(t)u|U(t)F(t)v\rangle ,$$

$$= \overline{\langle U(t)F(t)v|V(t)F(t)u\rangle} ,$$

$$= \overline{\langle S_F(t)v|u\rangle} = \langle u|S_F(t)v\rangle .$$

Thus $S_F(t)$ is self-adjoint.

If we pass to the limit in (1) using assumptions ii, iii, and iv we obtain

$$\langle UFu|VFu\rangle = \langle VFu|UFv\rangle .$$

Consequently there is a bounded self-adjoint operator S_F such that

$$\langle S_F u|v\rangle = \langle UFu|VFv\rangle .$$

Theorem 1a. Under the assumption i.-v. above if $z \in \underline{C}$, and if $\operatorname{Im} z \neq 0$, then

$$\{S_F(t) - zI\}^{-1} \longrightarrow \{S_F - zI\}^{-1} \quad \text{as} \quad t \to \infty .$$

Proof. It is enough to show that given $u,v \in \underset{\sim}{H}$ and any sequence $\underset{\sim}{P}_1$ of positive real numbers increasing to ∞, $\underset{\sim}{P}_1$ contains a subsequence $\underset{\sim}{P}_2$ such that

$$\langle \{S_F(t) - zI\}^{-1} u|v\rangle \rightarrow \langle \{S_F - zI\}^{-1} u|v\rangle \quad \text{as} \quad t \to \infty \quad \text{in} \quad \underset{\sim}{P}_2 .$$

Let us set

$$\{S_F(t) - zI\}^{-1}u = w(t) \qquad 0 < t < \infty .$$

Since the operators $\{S_F(t) - zI\}^{-1}$ are uniformly bounded, and since bounded sets in $\underset{\sim}{H}$ are conditionally compact in the weak topology, there exists a subsequence $\underset{\sim}{P}_2$ of $\underset{\sim}{P}_1$ and a $w \in \underset{\sim}{H}$ such that

$$(2) \qquad w(t) \rightharpoonup w \qquad \text{as } t \to \infty \text{ in } \underset{\sim}{P}_2 \quad .$$

It remains to identify w as $\{S_F - zI\}^{-1}u$. We have

$$S_F(t)w(t) = u + zw(t) \quad ,$$

from which we see that

$$(3) \qquad \langle S_F(t)w(t) | v\rangle \to \langle u + zw | v\rangle \quad \text{as } t \to \infty \text{ in } \underset{\sim}{P}_2 \ .$$

On the other hand

$$(4) \qquad \langle S_F(t)w(t) | v\rangle = \langle U(t)F(t)w(t) | V(t)F(t)v\rangle \quad .$$

It follows from assumption iii that

$$(5) \qquad V(t)F(t)v \to VFv \qquad \text{as } t \to \infty \ .$$

Assumption i and (2) imply that

$$(6) \qquad F(t)w(t) \rightharpoonup Fw \qquad \text{as } t \to \infty \text{ in } \underset{\sim}{P}_2 \ .$$

We also have

$$(7) \qquad \|U(t)[F(t)w(t)]\| = O(1) \qquad \text{as } t \to \infty \quad .$$

Assumption v. together with (6) and (7) implies that

(8) $$U(t)F(t)w(t) \longrightarrow UFw \quad \text{as} \quad t \to \infty \quad \text{in} \quad P_2 \quad .$$

See Lemma 2a of II. Inserting (5) and (8) in (4) we obtain

$$\langle S_F(t)w(t)|v\rangle \to \langle UFw|VFv\rangle = \langle S_Fw|v\rangle$$

as $t \to \infty$ in $\underset{\sim}{P}_2$. Together with (3) this gives

$$\langle S_Fw|v\rangle = \langle u+zw|v\rangle$$

which in turn implies that

$$\langle \{S_F - zI\}w|v\rangle = \langle u|v\rangle \quad .$$

Because this is valid for every $v \in \underset{\sim}{H}$

$$\{S_F - zI\}w = u, \text{ or } \quad w = \{S_F - zI\}^{-1}u \quad ,$$

and our proof is complete.

Theorem 1b. Let i. - v. hold and let

$$S_F = \int \lambda dC(\lambda), \; S_F(t) = \int \lambda dC(t,\lambda)$$

be the spectral resolutions of S_F and $S_F(t)$, normalized by the conditions $C(t,\lambda)$ and $C(\lambda)$ are right continuous, and $C(t,\infty) = C(\infty) = I$. Then

$$C(t,\lambda) \to C(\lambda) \qquad \text{as} \quad t \to \infty$$

for all $\lambda \notin \sigma_p(S_F)$, the point spectrum of S_F .

Proof. The arguments used to prove Theorem 3a of II apply here almost without change.

2. Convergence in Dimension

Let X and $X(t)$, $0 < t < \infty$, be uniformly bounded self-adjoint operators on $\underset{\sim}{H}$ and let

$$X = \int \lambda dC(\lambda) \quad \text{and} \quad X(t) = \int \lambda dC(t;\lambda)$$

be their spectral resolutions normalized as in Theorem 1b. Suppose that the following two assumptions hold.

a. $C(t;\lambda) \rightarrow C(\lambda)$ as $t \to \infty$ for all $\lambda \notin \sigma_p(X)$;

b. there is a number $m > 0$ such that if $u(t) \in \underset{\sim}{H}$, $\|u(t)\| = 1$, $0 < t < \infty$, and if

$$\langle X(t)u(t) | u(t)\rangle \geq m_1 > m \qquad t \in \underset{\sim}{P}_1 ,$$

where $\underset{\sim}{P}_1$ is some sequence $0 < t_1 < t_2 < \cdots$, $\lim t_k = \infty$, then $\underset{\sim}{P}_1$ contains a subsequence $\underset{\sim}{P}_2$ such that

$$u(t) \rightarrow u \neq 0 \qquad \text{as } t \to \infty \text{ in } P_2 .$$

Theorem 2a. Under assumptions a. and b. we have

(1) $$\dim[I - C(\lambda)] < \infty \qquad \text{for } \lambda > m ,$$

and

(2) $$\lim_{t \to \infty} \dim[I - C(t;\lambda)] = \dim[I - C]$$

for $\lambda > m$, $\lambda \notin \sigma_p(X)$.

Proof. The demonstration of this result is so much like that of Theorem 4a. of II that we omit it.

Theorem 2b. Let $X(t)$ and X satisfy assumptions a. and b., and let $Y(t)$ be a self-adjoint operator such that $Y(t) \rightrightarrows 0$ as $t \to +\infty$. Then $X(t) + Y(t)$ and X satisfy assumptions a. and b.

Proof. Let

$$X(t) + Y(t) = \int \lambda dC'(t;\lambda)$$

be the spectral resolution of $X(t) + Y(t)$. It follows easily from a. that

(3) $$\{X(t) - zI\}^{-1} \longrightarrow \{X - zI\}^{-1} \quad \text{as} \quad t \to \infty$$

for all z with $\operatorname{Im} z \neq 0$. We have

$$\{X(t) + Y(t) - zI\}^{-1} = \{X(t) - zI\}^{-1}\{I + Y(t)[X(t) - zI]^{-1}\}^{-1} .$$

If $z = x + iy$ then $\|\{X - zI\}^{-1}\| \leq y{-1}$. Thus

$$\{X(t) + Y(t) - zI\}^{-1} = \{X(t) - zI\}^{-1} \sum_{j=0}^{\infty} (-1)^j \{Y(t)[X(t) - zI]^{-1}\}^j ,$$

provided $\|Y(t)\| < |y|$. Letting $t \to \infty$ we see that the series

$$\sum_{j=0}^{\infty} (-1)^j \{Y(t)[X(t) - zI]^{-1}\}^j$$

converges to I in the uniform operator topology, and therefore

$$\{X(t) + Y(t) - zI\}^{-1} \longrightarrow \{X - zI\}^{-1} \quad \text{as} \quad t \to \infty .$$

The argument used to prove Theorem 3a in II can be applied once more to show that

$$C'(t,\lambda) \to C(\lambda) \qquad \text{as} \quad t \to \infty$$

for all λ not in the point spectrum of X.

Let $m_1 > m$ and suppose that for $0 < t < \infty$ $u(t) \in \underset{\sim}{H}$, $\|u(t)\| \leq 1$ and

$$\underline{\lim}_{t \to \infty} \langle [X(t) + Y(t)]u(t) | u(t) \rangle \geq m_1 > 0 .$$

Since $Y(t)u(t) \to 0$ as $t \to \infty$ it is clear that

$$\underline{\lim}_{t \to \infty} \langle X(t)u(t) | u(t) \rangle = \underline{\lim}_{t \to \infty} \langle [X(t) + Y(t)]u(t) | u(t) \rangle \geq m_1 .$$

Now use the fact $X(t)$ and X satisfy assumption b.

Corollary 2c. Under the assumptions of Theorem 2b we have:

$$C'(t;\lambda) \to C(\lambda) \qquad \lambda \notin \sigma_p(X);$$

$$\dim[I - C'(\lambda)] < \infty \qquad \lambda > m;$$

and, if $\lambda > m$ and is not in the point spectrum of X,

$$\lim_{t \to \infty} \dim[I - C'(t;\lambda)] = \dim[I - C(\lambda)] \qquad \lambda > m .$$

CHAPTER VI

THE FOURIER SERIES AND FOURIER TRANSFORM THEOREMS

1. Spaces and Operators

In all that follows we will make use of the Hilbert spaces $\underset{\sim}{E}$, $\underset{\sim}{E}^{\wedge}$, $\underset{\approx}{H}^{\wedge}$, and $\underset{\approx}{H}$, and the mappings between them defined in §1 of IV.

$$
\begin{array}{ccccc}
\underset{\sim}{E} & & & & \underset{\approx}{H} = \underset{\sim}{H} \oplus \cdots \oplus \underset{\sim}{H} \quad \text{(p summands)} \\
\varphi^{-1}\uparrow \;\; \downarrow\varphi & & & & \Psi^{-1}\uparrow \;\; \downarrow\Psi \\
\underset{\sim}{E}^{\wedge} & \xrightarrow{\;\;\underline{\chi}(t)\;\;} & & & \underset{\approx}{H}^{\wedge} = \underset{\sim}{H}^{\wedge} \oplus \cdots \oplus \underset{\sim}{H}^{\wedge} \quad \text{(p summands).} \\
 & \xleftarrow{\;\;\underline{\chi}(t)^{*}\;\;} & & &
\end{array}
$$

Following the conventions adopted there we use different independent variables to distinguish between functions on different spaces. In addition, functions on $\underset{\sim}{E}^{\wedge}$, $\underset{\sim}{H}^{\wedge}$ and $\underset{\approx}{H}^{\wedge}$ have "$^{\wedge}$'s" and functions on $\underset{\approx}{H}^{\wedge}$ and $\underset{\approx}{H}$ are in bold face. For example, $\underline{u}^{\wedge} = (u_1^{\wedge},\ldots,u_p^{\wedge})$ is in $\underset{\approx}{H}^{\wedge}$. If in an argument $v^{\wedge}(\underline{\theta})$ and $v(\underline{k})$ both appear then $v^{\wedge} = \varphi v$, and similarly if both $\underline{u}^{\wedge}(\underline{x})$ and $\underline{u}(\underline{y})$ appear then $\underline{u}^{\wedge} = \Psi\underline{u}$, etc .

Let $W(\underline{y})$ be a complex function on $\underline{R}_n$ belonging to $L^{\infty}(\underline{R}_n) \cap L^2(\underline{R}_n)$ which satisfies the following conditions.

(1)

a. $W(\underline{y})$ is locally Riemann integrable.

b. If $W^{\#}(\underline{y}) = \ell.u.b.\{|W(y-y')| : -\frac{1}{2} \le y_i' \le \frac{1}{2}\ ,\ i = 1,\ldots,m\}$

then $W^{\#} \in L^2(\underline{R}_n)$.

Note this implies that $W(\underline{y})$ is bounded and that $W(\underline{y}) \to 0$ as $\underline{y} \to \infty$.

Lemma 1a. If $t \geq 1$ then

$$\sum_{\underset{\sim}{Z}_n} |W(t^{-1}\underline{k})|^2 \leq t^n \int_{\underline{R}_n} W^{\#}(\underline{y})^2 d\underline{y} \quad .$$

Proof. Associate with each point $t^{-1}\underline{k}$ an n-dimensional cube $c(\underline{k},t)$ in $\underline{R}_n$ with sides of length t^{-1}, parallel to the axes, and centered on $t^{-1}\underline{k}$. Then

$$|W(t^{-1}\underline{k})|^2 \leq t^n \int_{c(\underline{k},t)} W^{\#}(\underline{y})^2 d\underline{y} \quad .$$

Summing over $\underline{k} \in \underline{Z}_n$ we obtain the desired conclusion.

We define an operator $E(t)$ on $\underset{\sim}{E}$ by the formula

$$E(t)u(\underline{k}) = W(t^{-1}\underline{k})u(\underline{k}) \quad .$$

It is apparent that if $0 < t < \infty$

$$\|E(t)u\|_{\underset{\sim}{E}} \leq \|W\|_\infty \|u\|_{\underset{\sim}{E}} \quad .$$

If $t \geq 1$ then

$$\|E(t)u(\underline{k})\|_1 = \sum_{\underline{Z}_n} |W(t^{-1}\underline{k})u(\underline{k})| \quad ,$$

$$\leq \left\{ \sum_{\underline{Z}_n} |W(t^{-1}\underline{k})|^2 \right\} \|u(\underline{k})\|_{\underset{\sim}{E}} \quad ,$$

$$\leq t^{n/2} \|W^{\#}(\underline{y})\|_2 \|u(\underline{k})\|_{\underset{\sim}{E}} \quad .$$

by Lemma 1a.

Let us set

$$E^{\wedge}(t)u^{\wedge}\cdot(\underset{\sim}{\theta}) = [\varphi\, E(t)\varphi^{-1}u^{\wedge}]\cdot(\underset{\sim}{\theta}) \qquad u^{\wedge} \in \underset{\sim}{E}^{\wedge} .$$

Clearly we have

$$\|E^{\wedge}(t)u^{\wedge}(\underset{\sim}{\theta})\|_{\underset{\sim}{E}^{\wedge}} \leq \|W\|_{\infty}\, \|u^{\wedge}\|_{\underset{\sim}{E}^{\wedge}} \quad , \tag{2}$$

and

$$\|E^{\wedge}(t)u^{\wedge}(\underset{\sim}{\theta})\|_{\infty} \leq t^{n/2}\|W^{\#}\|_2\, \|u^{\wedge}\|_{\underset{\sim}{E}^{\wedge}} \qquad \text{if} \quad t \geq 1 . \tag{3}$$

For $f^{\wedge}(\underset{\sim}{\theta})$ any real function in $L^1(\underset{\sim}{T}_n)$ consider the sesquilinear form

$$(u^{\wedge},v^{\wedge})_t = (2\pi)^{-n}\int_{\underset{\sim}{T}_n} f^{\wedge}(\underset{\sim}{\theta})[E^{\wedge}(t)u^{\wedge}\cdot(\underset{\sim}{\theta})][\overline{E^{\wedge}(t)v^{\wedge}\cdot(\underset{\sim}{\theta})}]d\underset{\sim}{\theta} \quad .$$

It follows from (3) that $(u^{\wedge},v^{\wedge})_t$ is bounded for each $t > 0$. Since $(u^{\wedge},v^{\wedge})_t$ is Hermitian symmetric,

$$(u^{\wedge},v^{\wedge})_t = \overline{(v^{\wedge},u^{\wedge})}_t \quad ,$$

there exists a bounded self-adjoint operator $T_E^{\wedge}(t)$ on $\underset{\sim}{E}^{\wedge}$ such that

$$\langle T_E^{\wedge}(t)u^{\wedge}|v^{\wedge}\rangle_{\underset{\sim}{E}^{\wedge}} = (2\pi)^{-n}\int_{\underset{\sim}{T}_n} f^{\wedge}(\underset{\sim}{\theta})[E^{\wedge}(t)u^{\wedge}\cdot(\underset{\sim}{\theta})][\overline{E^{\wedge}(t)v^{\wedge}\cdot(\underset{\sim}{\theta})}]d\underset{\sim}{\theta} . \tag{4}$$

Our goal is to describe the largest positive and largest negative eigen values of $T_E^{\wedge}(t)$ as $t \to +\infty$. Such a description is possible by our methods only if $f^{\wedge}(\underset{\sim}{\theta})$ has the structure we will now describe.

Let us fix once and for all a real number ω, $0 < \omega < n$. We denote by $L(t)$ a fixed positive continuous function on $(0,\infty)$ which is slowly oscillating at 0. (For reasons of convenience we assume $L(t)$ is 1 for all sufficiently large t.) It is helpful to require that $L(t)$ have a further property, which could be dispensed with at the cost of a somewhat more complicated formulation. We assume that if $r(t) > 0$ for $0 < t < \infty$, and if

(5) $$\log r(t) = o[\log t^{-1}] \qquad \text{as } t \to \infty+ ,$$

then

(6) $$L(t^{-1}r(t)) \sim L(t^{-1}) \qquad \text{as } t \to \infty+ .$$

We fix two integers $0 < q < p$, and p distinct points $\{\underline{\theta}_k\}_1^p$ in $\underline{T}_n$. In what follows we will identify $\underline{T}_n$ with the cube $-\pi \leq \theta_j < \pi$ $j = 1,\ldots,n$. We can, without loss of generality, assume that no $\underline{\theta}_k$ lies on the boundary of this cube. Let

$$\underline{T}_n = \bigcup_{k=1}^{p} R_k$$

be a decomposition of $\underline{T}_n$ into non-overlapping rectangles with sides parallel to the coordinate axes such that $\underline{\theta}_k$ lies in the interior of R_k for $1 \leq k \leq p$. For each k let Φ_k be a non-negative bounded measurable function on $\underline{R}_n$, homogeneous of degree 0, but not zero almost everywhere. Finally let $\gamma(k)$ be $+1$ for $1 \leq k \leq q$ and -1 for $q < k \leq p$.

We wish $f^{\wedge}(\underline{\theta})$ to behave like

$$\gamma(k)|\underline{\theta}-\underline{\theta}_k|^{-\omega} L(|\underline{\theta}-\underline{\theta}_k|)^{-1} \phi(\underline{\theta}-\underline{\theta}_k)$$

near $\underline{\theta}_k$ $k = 1,\ldots,p$, and to relatively well behaved elsewhere. However, it is convenient to make this precise in a way which differs from that used in the parallel situation of Chapters III and IV.

Definition 1b. A real function $h^{\wedge}(\underline{\theta})$ on $\underline{T}_n$ is said to be "negligible" with respect to ω and L if

(8) $$\int_{A(\tau)} |h^{\wedge}(\underline{\theta})|\,d\underline{\theta} = o\{L(\tau^{-1/\omega})^{-n/\omega}\tau^{(\omega-n)/\omega}\} \quad \text{as} \quad \tau \to \infty$$

where

$$A(\tau) = \{\underline{\theta} \in \underline{T}_n : |h^{\wedge}(\underline{\theta})| > \tau\} .$$

It is evident that if $h^{\wedge}(\underline{\theta})$ is negligible it is à fortiori in $L^1(\underline{T}_n)$.

We define $g^{\wedge}(\underline{\theta})$ on $\underline{T}_n$ by the formulas

(9) $$g^{\wedge}(\underline{\theta}) = \gamma(k)g_k^{\wedge}(\underline{\theta}) \qquad \underline{\theta} \in R_k \quad k = 1,\ldots,p$$

where

$$g_k^{\wedge}(\underline{\theta}) = |\underline{\theta}-\underline{\theta}_k|^{-\omega} L(|\underline{\theta}-\underline{\theta}_k|)^{-1} \phi_k(\underline{\theta}-\underline{\theta}_k) .$$

This definition is ambiguous if $\underline{\theta} \in R_k \cap R_j$ where $j \neq k$; however, such points form a set of measure 0. It is apparent that $g^{\wedge}(\underline{\theta}) \in L^1(\underline{T}_n)$.

In the remainder of this chapter we will suppose that

$$f^{\wedge}(\underline{\theta}) = g^{\wedge}(\underline{\theta}) + h^{\wedge}(\underline{\theta})$$

where $g^{\wedge}(\underline{\theta})$ is as in (9) and $h^{\wedge}(\underline{\theta})$ is negligible. Clearly $f^{\wedge}(\underline{\theta}) \in L^1(\underline{T}_n)$. $T_E^{\wedge}(t)$ is throughout the operator defined using this $f^{\wedge}(\underline{\theta})$.

2. Further Operators

We need a decomposition of $T_E^{\wedge}$ corresponding to this decomposition of $f^{\wedge}(\underline{\theta})$. Consider the sesquilinear forms on $\underset{\sim}{E}^{\wedge}$

$$(u^{\wedge},v^{\wedge})_t' = (2\pi)^{-n} \int_{\underline{T}_n} g^{\wedge}(\underline{\theta})[E^{\wedge}(t)u^{\wedge}\cdot(\underline{\theta})]\overline{[E^{\wedge}(t)v^{\wedge}\cdot(\underline{\theta})]}d\underline{\theta} \quad ,$$

and

$$(u^{\wedge},v^{\wedge})_t'' = (2\pi)^{-n} \int_{\underline{T}_n} h^{\wedge}(\underline{\theta})[E^{\wedge}(t)u^{\wedge}\cdot(\underline{\theta})]\overline{[E^{\wedge}(t)v^{\wedge}\cdot(\underline{\theta})]}d\underline{\theta} \quad .$$

It is easy to verify that these forms are bounded and symmetric so that there exist bounded self-adjoint operators $S_E^{\wedge}(t)$ and $N_E^{\wedge}(t)$ such that

$$\langle S_E^{\wedge}(t)u^{\wedge}|v^{\wedge}\rangle_{\underset{\sim}{E}^{\wedge}} = (u^{\wedge},v^{\wedge})_t' \quad ,$$

and

$$\langle N_E^{\wedge}(t)u^{\wedge}|v^{\wedge}\rangle_{\underset{\sim}{E}^{\wedge}} = (u^{\wedge},v^{\wedge})_t'' \quad .$$

Clearly we have

$$T_E^{\wedge}(t) = S_E^{\wedge}(t) + N_E^{\wedge}(t) \quad .$$

In order to apply the results of Chapter VI we introduce the unbounded operators

$$(1)\qquad \begin{aligned} U^{\wedge}u^{\wedge}\cdot(\underline{\theta}) &= |g^{\wedge}(\underline{\theta})|^{\frac{1}{2}}\, u^{\wedge}(\underline{\theta}) \quad ,\\ V^{\wedge}u^{\wedge}\cdot(\underline{\theta}) &= g^{\wedge}(\underline{\theta})|g^{\wedge}(\underline{\theta})|^{-\frac{1}{2}}\, u^{\wedge}(\theta) \quad , \end{aligned} \qquad u^{\wedge}\in \underset{\sim}{E}^{\wedge}$$

where $\underset{\sim}{D}[U^{\wedge}] = \underset{\sim}{D}[V^{\wedge}]$ consists of all functions $u^{\wedge}(\theta)\in \underset{\sim}{E}^{\wedge}$ such that $|g^{\wedge}(\underline{\theta})|^{\frac{1}{2}}\, u^{\wedge}(\underline{\theta}) \in L^2(\underline{T}_n)$. Since $\underset{\sim}{R}[E^{\wedge}(t)] \subset L^{\infty}(\underline{T}_n) \subset \underset{\sim}{D}[U^{\wedge}] = \underset{\sim}{D}[V^{\wedge}]$ we see that $U^{\wedge}E^{\wedge}(t)$ and $V^{\wedge}E^{\wedge}(t)$ are bounded operators. We now have the formulas

$$\begin{aligned} \langle S_E^{\wedge}(t)u^{\wedge}|v^{\wedge}\rangle_{\underset{\sim}{E}^{\wedge}} &= \langle U^{\wedge}E^{\wedge}(t)u^{\wedge}|V^{\wedge}E^{\wedge}(t)v^{\wedge}\rangle_{\underset{\sim}{E}^{\wedge}} \quad ,\\ &= \langle V^{\wedge}E^{\wedge}(t)u^{\wedge}|U^{\wedge}E^{\wedge}(t)v^{\wedge}\rangle_{E^{\wedge}} \quad . \end{aligned}$$

As before, the perturbation argument which yields our final result is carried out in $\underset{\approx}{H}^{\wedge}$ Let $\underline{\chi}(t)$ be as in §1 of Chapter IV; then

$$\begin{aligned} \underline{F}^{\wedge}(t) &= \underline{\chi}(t)E^{\wedge}(t)\underline{\chi}(t)^{*} \quad ,\\ \underline{T}_{\underline{F}}^{\wedge}(t) &= t^{-\omega}\, L(t^{-1})\underline{\chi}(t)T_E^{\wedge}(t)\underline{\chi}(t)^{*} \quad ,\\ \underline{S}_{\underline{F}}^{\wedge}(t) &= t^{-\omega}\, L(t^{-1})\underline{\chi}(t)S_E^{\wedge}(t)\underline{\chi}(t)^{*} \quad ,\\ \underline{N}_{\underline{F}}^{\wedge}(t) &= \left[t^{-\omega}L(t^{-1})\right]^{\frac{1}{2}}\, \underline{\chi}(t)N_E^{\wedge}(t)\underline{\chi}(t)^{*} \quad ,\\ \underline{U}^{\wedge}(t) &= \left[t^{-\omega}L(t^{-1})\right]^{\frac{1}{2}}\, \underline{\chi}(t)U^{\wedge}\underline{\chi}(t)^{*} \quad ,\\ \underline{V}^{\wedge}(t) &= \left[t^{-\omega}L(t^{-1})\right]^{\frac{1}{2}}\, \underline{\chi}(t)V^{\wedge}(t)\underline{\chi}(t)^{*} \quad . \end{aligned}$$

Note that $\underline{T}_F^{\wedge}(t)$ is the direct sum of an operator unitarily equivalent to $\underline{T}_E^{\wedge}$ and a null operator. We further note that

$$\underline{T}_F^{\wedge}(t) = \underline{S}_F^{\wedge}(t) + \underline{N}_F^{\wedge}(t)$$

and that

$$\langle \underline{S}_F^{\wedge}(t)\underline{u}^{\wedge} | \underline{v}^{\wedge} \rangle_{\underset{\sim}{\underline{H}}^{\wedge}} = \langle \underline{U}^{\wedge}(t)\underline{F}^{\wedge}(t)\underline{u}^{\wedge} | \underline{V}^{\wedge}(t)\underline{F}^{\wedge}(t)\underline{v}^{\wedge} \rangle_{\underset{\sim}{\underline{H}}^{\wedge}} ,$$

$$= \langle \underline{V}^{\wedge}(t)\underline{F}^{\wedge}(t)\underline{u}^{\wedge} | \underline{U}^{\wedge}(t)\underline{F}^{\wedge}(t)\underline{v}^{\wedge} \rangle_{\underset{\sim}{\underline{H}}^{\wedge}} .$$

We conclude by defining the limit operators of our perturbation theory. Let

$$Fu\cdot(\underline{y}) = W(\underline{y})u(\underline{y}) \qquad u \in \underset{\sim}{H} ,$$

$$F^{\wedge}u^{\wedge} = \Phi F \Phi^{-1} u^{\wedge} \qquad u^{\wedge} \in \underset{\sim}{H}^{\wedge} ,$$

and

$$\underline{F}^{\wedge}\underline{u}^{\wedge} = (F^{\wedge}u_1^{\wedge}, \ldots, F^{\wedge}u_p^{\wedge}) \qquad \underline{u}^{\wedge} \in \underset{\sim}{\underline{H}}^{\wedge} .$$

For $k = 1,\ldots,p$ let

$$G(k,\underline{x}) = |\underline{x}|^{-(1)} \Phi_k(\underline{x}) ,$$

and define the unbounded operators

$$\left.\begin{array}{l} U^{\wedge}(k)u^{\wedge}(\underline{x}) = G(k,\underline{x})^{\frac{1}{2}} u^{\wedge}(\underline{x}) \\ V^{\wedge}(k)u^{\wedge}(\underline{x}) = \gamma(k)G(k,\underline{x})^{\frac{1}{2}} u^{\wedge}(\underline{x}) \end{array}\right\} \quad u^{\wedge} \in \underset{\sim}{H}^{\wedge} ,$$

where $u^{\wedge} \in \underset{\sim}{D}[U^{\wedge}(k)] \equiv \underset{\sim}{D}[V^{\wedge}(k)]$ if and only if $|u^{\wedge}(x)|^2 G(k,\underline{x}) \in L^1(\underline{R}_n)$.

We set

$$\underline{U}^\wedge \underline{u}^\wedge = (U^\wedge(1)u_1^\wedge,\ldots,U^\wedge(p)u_p^\wedge)$$

$$\underline{u}^\wedge \in \underset{\approx}{H}^\wedge \quad .$$

$$\underline{V}^\wedge \underline{v}^\wedge = (V^\wedge(1)u_1^\wedge,\ldots,V^\wedge(p)v_p^\wedge)$$

Since

(2) $$\|F^\wedge u^\wedge\|_{\underset{\sim}{H}^\wedge} \leq \|W\|_\infty \, \|u^\wedge\|_{\underset{\sim}{H}^\wedge}$$

and

(3) $$\|F^\wedge u^\wedge\|_\infty \leq \|W\|_2 \, \|u^\wedge\|_{\underset{\sim}{H}^\wedge}$$

we see that $\underset{\sim}{R}(F^\wedge) \subset \underset{\sim}{D}[U^\wedge(k)] \equiv \underset{\sim}{D}[V^\wedge(k)]$ for $k = 1,\ldots,p$. Consequently

$$\langle \underline{U}^\wedge \underline{F}^\wedge \underline{u}^\wedge | \underline{V}^\wedge \underline{F}^\wedge \underline{v}^\wedge \rangle_{\underset{\approx}{H}^\wedge} = \langle \underline{S}_{\underline{F}}^\wedge \, \underline{u}^\wedge | \underline{v}^\wedge \rangle_{\underset{\approx}{H}^\wedge}$$

defines $\underline{S}_{\underline{F}}^\wedge$ as a bounded linear operator on $\underset{\approx}{H}^\wedge$, and, since

$$\langle \underline{U}^\wedge \underline{F}^\wedge \underline{u}^\wedge | \underline{V}^\wedge \underline{F}^\wedge \underline{v}^\wedge \rangle_{\underset{\approx}{H}^\wedge} = \langle \underline{V}^\wedge \underline{F}^\wedge \underline{u}^\wedge | U^\wedge F^\wedge v^\wedge \rangle_{\underset{\approx}{H}^\wedge}$$

$\underline{S}_{\underline{F}}^\wedge$ is self-adjoint.

3. $\underline{F}^\wedge(t)$ and $\underline{F}^\wedge(t)^*$

Since $\underset{\approx}{H}^\wedge = H^\wedge \oplus \cdots \oplus H^\wedge$ (p summands) it follows that $\underline{F}^\wedge = \underline{\chi}(t)E^\wedge(t)\underline{\chi}(t)^*$ can be represented as a $p \times p$ matrix of operators

$$[F^\wedge(t,j,k)] \qquad j,k = 1,\ldots,p \quad ,$$

where $F^{\wedge}(t,j,k)$ maps the k-th summand of $\underset{\approx}{H}^{\wedge}$ into the j-th summand of $\underset{\approx}{H}^{\wedge}$. Alternatively, since all the summands of $\underset{\approx}{H}^{\wedge}$ are copies of $\underset{\sim}{H}^{\wedge}$ we can, if we wish, study $F^{\wedge}(t,j,k)$ as a mapping of $\underset{\sim}{H}^{\wedge}$ into itself.

Lemma 3a. Let $u^{\wedge} \in \underset{\sim}{H}^{\wedge}$; then $F^{\wedge}(t,j,k)u^{\wedge}(\underline{x})$ is equal to

$$t^{-n} \sum_{\underline{m} \in \underline{Z}_n} e^{it^{-1}\underline{m}\cdot(\underline{x}+t\underline{\theta}_j)} W(t^{-1}\underline{m})(2\pi)^{-n} \int_{R_k^t} u^{\wedge}(\underline{x}')e^{-it^{-1}\underline{m}\cdot(\underline{x}'+t\underline{\theta}_k)}d\underline{x}'$$

if $\underline{x} \in R_j^t$, and is equal to 0 if $\underline{x} \notin R_j^t$.

Proof. This follows from a short computation analogous to the one in §2 of IV.

Theorem 3b. If $u^{\wedge} \in \underset{\sim}{H}^{\wedge}$ then for $t \geq 1$ we have

$$(1) \qquad \|F^{\wedge}(t,j,k)u^{\wedge}\|_{\underset{\sim}{H}^{\wedge}} \leq \|W\|_{\infty} \|u^{\wedge}\|_{\underset{\sim}{H}^{\wedge}} ,$$

$$(2) \qquad \|F^{\wedge}(t,j,k)u^{\wedge}\|_{\infty} \leq \|W^{\#}\|_2 \|u^{\wedge}\|_{\underset{\sim}{H}^{\wedge}} .$$

Proof. Since $\underline{\chi}(t)$ is isometric and $\underline{\chi}(t)^*$ is partially isometric it follows from (2) of §1 that

$$\|\underline{F}^{\wedge}(t)\underline{u}^{\wedge}\|_{\underset{\approx}{H}^{\wedge}} \leq \|W\|_{\infty} \|\underline{u}^{\wedge}\|_{\underset{\approx}{H}^{\wedge}} ,$$

which is stronger result than the first inequality in (1).

To demonstrate (2) we proceed directly. If

$$u(t,\underline{m}) = (2\pi)^{-n} \int_{R_k^t} u^\wedge(\underline{x}')e^{-it^{-1}\underline{m}\cdot\underline{x}'}d\underline{x}'$$

then, by Parseval's equality,

$$\sum_{\underline{m}} |u(t,\underline{m})|^2 = t^n(2\pi)^{-n} \int_{R_k^t} |u^\wedge(\underline{x}')|^2 d\underline{x}' \leq t^n \|u^\wedge\|_{H^\wedge}^2 \quad .$$

Applying Lemma 3a, Schwarz's inequality, and finally Lemma 1a we obtain

$$|F^\wedge(t,j,k)u^\wedge\cdot(\underline{x})| \leq t^{-n} \sum_{\underline{m}} |W(t^{-1}\underline{m})| \, |u(t,\underline{m})| \quad ,$$

$$\leq \|W^\#\|_2 \, \|u^\wedge\|_{\underset{\sim}{H}^\wedge} \quad .$$

Theorem 3c. For every $\underline{u} \in \underset{\approx}{H}^\wedge$ we have

(3) $$\|[\underline{F}^\wedge(t) - \underline{F}^\wedge]\underline{u}^\wedge\|_{\underset{\approx}{H}^\wedge} \to 0 \quad \text{as} \quad t \to \infty \quad ,$$

$$\|[\underline{F}^\wedge(t) - \underline{F}^\wedge]\underline{u}^\wedge\|_\infty \to 0 \quad \text{as} \quad t \to \infty \quad ,$$

and

(4) $$\|[\underline{F}^\wedge(t)^* - \underline{F}^{\wedge *}]\underline{u}^\wedge\|_{\underset{\approx}{H}^\wedge} \to 0 \quad \text{as} \quad t \to \infty \quad ,$$

$$\|[\underline{F}^\wedge(t)^* - \underline{F}^{\wedge *}]\underline{u}^\wedge\|_\infty \to 0 \quad \text{as} \quad t \to \infty \quad .$$

Proof. If we can show that for j, k = 1,...,p we have

$$\|\{F^\wedge(t,j,k) - \delta(j,k)F^\wedge\}u^\wedge\|_{\underset{\sim}{H}^\wedge} \to 0 \quad \text{as} \quad t \to \infty \quad ,$$

and

$$\|\{F^\wedge(t,j,k) - \delta(j,k)F^\wedge\}u^\wedge\|_\infty \to 0 \quad \text{as} \quad t \to \infty \quad ,$$

for all $u^\wedge \in \underset{\sim}{H}^\wedge$, then (3) will follow. Let

$$W_1(\underline{y}) = X_\Omega(\underline{y})$$

where $X_\Omega(\underline{y})$ is the characteristic function of a set Ω of (finite)

Jordan content in $\underline{R}_n$. Let $F_1^\wedge(t,k,j)$ and $F_1^\wedge$ be constructed from $W_1(\underline{y})$

in the same way that $F^\wedge(t,k,j)$ and $F^\wedge$ are constructed from $W(\underline{y})$. We

assert that for all $u^\wedge \in H^\wedge$

(5) $$\|\{F_1^\wedge(t,j,k) - \delta(j,k)F_1^\wedge\}u^\wedge\|_{\underset{\sim}{H}^\wedge} \to 0 \quad \text{as} \quad t \to \infty \quad ,$$

and

(6) $$\|\{F_1^\wedge(t,j,k) - \delta(j,k)F_1^\wedge\}u'\|_\infty \to 0 \quad \text{as} \quad t \to \infty \quad .$$

Indeed the assertions (5) and (6) follow for all $u^\wedge \in \underset{\sim}{D}^\wedge(\Omega)$ and all

$u^\wedge \in \underset{\sim}{D}^\wedge(\Omega')$ from Theorem 4b of IV. Since $\partial\Omega$ has Lebesgue measure 0,

$\underset{\sim}{D}^\wedge(\Omega) + \underset{\sim}{D}^\wedge(\Omega')$ is dense in $\underset{\sim}{H}^\wedge$. Using (1) and (2), an evident approxima-

tion argument yields (3) and (4) for all $u^\wedge \in \underset{\sim}{H}^\wedge$. Let

$$W_2(\underline{y}) = \sum_{\mu=1}^{m} c(\mu)X_{\Omega_\mu}(\underline{y})$$

where each Ω_μ is the characteristic function of a set of Jordan content and each $c(\mu)$ is a complex constant. Let $F_2^\wedge(t,j,k)$ and $F_2^\wedge$ be constructed from $W_2(\underline{y})$ in the same way that $F^\wedge(t,j,k)$ and $F^\wedge$ are constructed from $W(\underline{y})$. It follows immediately from (5) and (6) that

(7) $$\|\{F_2^\wedge(t,j,k) - \delta(j,k)F_2^\wedge\}u^\wedge\|_{\underset{\sim}{H}^\wedge} \rightarrow 0 \quad \text{as } t \to \infty \quad ,$$

and

(8) $$\|\{F_2^\wedge(t,j,k) - \delta(j,k)F_2^\wedge\}u^\wedge\|_\infty \rightarrow 0 \quad \text{as} \quad t \to \infty$$

for all $u^\wedge \in \underset{\sim}{H}^\wedge$.

Let W be as in §1. Given $\epsilon > 0$ choose a $W_2(\underline{y})$ such that

$$\|(W-W_1)\cdot(\underline{y})\|_\infty < \epsilon \ , \ \|(W-W_2)^{\#}\cdot(\underline{y})\|_2 < \epsilon \quad .$$

Then by Theorem 3b applied to $W(\underline{y}) - W_2(\underline{y})$ we have for $t \geq 1$

$$\|\{F^\wedge(t,j,k) - F_2^\wedge(t,j,k)\}u^\wedge\|_{\underset{\sim}{H}^\wedge} \leq \epsilon\|u^\wedge\|_{\underset{\sim}{H}^\wedge} \quad ,$$

$$\|\{F^\wedge(t,j,k) - F_2^\wedge(t,j,k)\}u^\wedge\|_\infty \leq \epsilon\|u^\wedge\|_{\underset{\sim}{H}^\wedge} \quad .$$

We also have as a corollary that

$$\|\{F^\wedge-F_2^\wedge\}u^\wedge\|_{\underset{\sim}{H}^\wedge} < \epsilon\|u^\wedge\|_{\underset{\sim}{H}^\wedge} \quad , \ \|\{F^\wedge - F_2^\wedge\}u^\wedge\|_\infty < \epsilon\|u^\wedge\|_{\underset{\sim}{H}^\wedge} \ ,$$

for all $u^\wedge \in \underset{\sim}{H}^\wedge$. In conjunction with (7) and (8) these imply

$$\overline{\lim_{t \to \infty}} \|\{F^\wedge(t,j,k) - \delta(j,k)F^\wedge\}u^\wedge\|_{\underset{\sim}{H}^\wedge} \leq 2\epsilon\|u^\wedge\|_{\underset{\sim}{H}^\wedge} ,$$

and

$$\overline{\lim_{t \to \infty}} \|\{F^\wedge(t,j,k) - \delta(j,k)F^\wedge\}u^\wedge\|_\infty \leq 2\epsilon\|u^\wedge\|_{\underset{\sim}{H}^\wedge} .$$

Since ϵ is arbitrary the proof of (3) is complete.

We have

$$\underline{F}^\wedge(t)^* = \underline{X}(t)E(t)^*\underline{X}(t)^*$$

where $E^*(t)$ is constructed from $\overline{W(\underline{y})}$ in exactly the same way that $E(t)$ is constructed from $W(\underline{y})$. Thus we have also proved (4).

4. $\underline{U}^\wedge(t)\underline{F}^\wedge(t)$ and $\underline{U}^\wedge\underline{F}^\wedge$, $\underline{V}^\wedge(t)\underline{F}^\wedge(t)$ and $\underline{V}^\wedge\underline{F}^\wedge$.

It is obvious from the definitions of the operators involved that $\underset{\sim}{R}[\underline{F}^\wedge(t)] \subset \underset{\sim}{D}[\underline{U}^\wedge(t)] = \underset{\sim}{D}[\underline{V}^\wedge(t)]$ so that $\underline{U}^\wedge(t)\underline{F}^\wedge(t)$ and $\underline{V}^\wedge(t)\underline{F}^\wedge(t)$ are bounded operators on $\underset{\approx}{H}^\wedge$. Let

$$G(t,m,\underline{x}) = \begin{cases} L(t^{-1})L(t^{-1}|\underline{x}|)^{-1}|\underline{x}|^{-\omega}\Phi_m(\underline{x}), & \underline{x} \in R_m^t \\ 0 & , \underline{x} \notin R_m^t \end{cases}$$

where $1 \leq m \leq p$ and let

$$U^\wedge(t,m)u^\wedge(\underline{x}) = G(t,m,\underline{x})^{\frac{1}{2}}u^\wedge(\underline{x}) ,$$

$$V^\wedge(t,m)u^\wedge(\underline{x}) = \gamma(m)G(t,m,\underline{x})^{\frac{1}{2}}u^\wedge(\underline{x}) , \qquad u^\wedge \in \underset{\sim}{H}^\wedge .$$

It can be immediately verified that

$$\underline{U}^{\wedge}(t)\underline{u}^{\wedge} = (U^{\wedge}(t,1)u_1^{\wedge},\ldots,U^{\wedge}(t,p)u_p^{\wedge}) \quad ,$$
$$\underline{V}^{\wedge}(t)\underline{u}^{\wedge} = (V^{\wedge}(t,1)v_1^{\wedge},\ldots,V^{\wedge}(t,p)v_p^{\wedge}) \quad . \qquad \underline{u}^{\wedge} \in \underline{\underline{H}}^{\wedge}$$

Fix $0 < \omega'' < \omega < \omega' < n$. If M is chosen sufficiently large and if

(1) $$G(\underline{x}) = \begin{cases} M|\underline{x}|^{\omega'} & |\underline{x}| \le 1 \\ M|\underline{x}|^{-\omega''} & |\underline{x}| > 1 \end{cases}$$

then

(2) $$G(m,\underline{x}) \le G(\underline{x}) \qquad \underline{x} \in \underline{R}_n \quad m = 1,\ldots,p$$

and

(3) $$G(t,m,\underline{x}) < G(\underline{x}) \qquad \underline{x} \in R_m^t,\ t \ge 1,\ m = 1,\ldots,p \ .$$

$G(\underline{x})$ plays a role as a majorant in various arguments which follow.

Theorem 42. As $t \to \infty$ we have

$$\underline{U}^{\wedge}(t)\underline{F}^{\wedge}(t) \to \underline{U}^{\wedge}\underline{F}^{\wedge} \quad ,$$
$$\underline{V}^{\wedge}(t)\underline{F}^{\wedge}(t) \to \underline{V}^{\wedge}\underline{F}^{\wedge} \quad .$$

Proof. In order to prove our first assertion it is enough to check that for $j, k = 1,\ldots,p$ we have

$$U^{\wedge}(t,j)F^{\wedge}(t,jk) - \delta(j,k)U^{\wedge}(j)F^{\wedge} \to 0$$

as $t \to \infty$. This in turn will follow if we can establish that as $t \to \infty$

and

$$U^{\wedge}(t,j)F^{\wedge}(t,j,k) - \delta(j,k)U^{\wedge}(t,j)F^{\wedge} \to 0 \quad ,$$

$$U^{\wedge}(t,j)F^{\wedge} - U^{\wedge}(j)F^{\wedge} \to 0.$$

If

$$I = \|U^{\wedge}(t,j)[F^{\wedge}(t,jk) - \delta(j,k)F^{\wedge}]u^{\wedge}\|^2_{\underset{\sim}{H}^{\wedge}}$$

then $I = I_1 + I_2$ corresponding to the ranges of integration $|\underline{x}| \leq 1$ and $|\underline{x}| > 1$. We have

$$I_1 \leq \|[F^{\wedge}(t,j,k) - \delta(j,k)F^{\wedge}]u^{\wedge}\|^2_{\infty}(2\pi)^{-n} \int_{|\underline{x}|\leq 1} 4G(\underline{x})d\underline{x} \quad ,$$

and

$$I_2 \leq \|[F^{\wedge}(t,j,k) - \delta(j,k)F^{\wedge}]u^{\wedge}\|^2_{\underset{\sim}{H}^{\wedge}}(4M) \quad .$$

By Theorem 3c both of these go to zero as t goes to infinity.

It remains to show that if

$$J = \|U^{\wedge}(t,j) - U^{\wedge}(j)]F^{\wedge}u^{\wedge}\|^2_{\underset{\sim}{H}^{\wedge}}$$

then $J \to 0$ as $t \to \infty$. We have

$$J = (2\pi)^{-n} \int_{\underline{R}_n} |G(t,j,\underline{x})^{\frac{1}{2}} - G(j,\underline{x})^{\frac{1}{2}}|^2 \, |F^{\wedge}u^{\wedge}(\underline{x})|^2 d\underline{x} = J_1 + J_2$$

corresponding to the ranges of integration $|\underline{x}| \leq L$ and $|\underline{x}| \geq 1$. By (2) of §3 and (3) of §2 the integrand in J_1 is dominated by

$$4G(\underline{x})\|W^{\#}\|_2^2 \, \|u^{\wedge}\|^2_{\underset{\sim}{H}^{\wedge}} \quad .$$

Similarly the integrand in J_2 is dominated by

$$4G|F^{\wedge}u^{\wedge} \cdot (\underline{x})|^2 \quad ,$$

both of which are integrable. Since

$$\lim_{t \to \infty} G(t,j,\underline{x})^{\frac{1}{2}} = G(j,x)^{\frac{1}{2}}$$

for all $\underline{x} \in \underline{R}_n$, the Lebesgue dominated convergence theorem can be applied to show that J_1 and J_2 go to zero as t goes to infinity. We have thus proved our first assertion. The second is proved in the same way.

5. $\underline{U}^{\wedge}(t)$ and $\underline{U}^{\wedge}$, $\underline{V}^{\wedge}(t)$ and $\underline{V}^{\wedge}$

Theorem 5a. $\underline{U}^{\wedge}$ is the closure of the strong limit of $\underline{U}^{\wedge}(t)$ as $t \to \infty$, and $\underline{V}^{\wedge}$ is the closure of the strong limit of $\underline{V}^{\wedge}(t)$ as $t \to \infty$.

Proof. Let $u^{\wedge} \in \underset{\sim}{D}[U^{\wedge}(j)]$. We must show that given $\epsilon > 0$ there exists $u_{\epsilon}^{\wedge} \in \underset{\sim}{H}^{\wedge}$ such that for each $j = 1,\ldots,n$

(1) $$U^{\wedge}(t,j)u_{\epsilon}^{\wedge} \to U^{\wedge}(j)u_{\epsilon}^{\wedge} \qquad \text{as} \quad t \to \infty \text{ m}$$

and such that

(2) $$\|u^{\wedge} - u_{\epsilon}^{\wedge}\|_{\underset{\sim}{H}^{\wedge}} < \epsilon \; , \; \|U^{\wedge}(j)(u^{\wedge} - u_{\epsilon}^{\wedge})\|_{\underset{\sim}{H}^{\wedge}} < \epsilon \; .$$

Given $u^{\wedge} \in \underset{\sim}{D}[U^{\wedge}(j)]$ let

$$u_T^{\wedge}(\underline{x}) = \begin{cases} u^{\wedge}(\underline{x}), \text{ if } T^{-1} < |\underline{x}| < T \\ 0 \quad , \text{ otherwise} \end{cases} ,$$

where $1 < T < \infty$. Using $|u^{\wedge}(\underline{x})|^2$ and $|U^{\wedge}(j)u^{\wedge}\cdot(x)|^2$ as majorants we can apply the Lebesgue dominated convergence theorem to show that

$$\lim_{T \to \infty} \|u^{\wedge} - u_T^{\wedge}\|_{\underset{\sim}{H}^{\wedge}} = \lim_{T \to \infty} \|U^{\wedge}(j)(u^{\wedge} - u_T^{\wedge})\|_{\underset{\sim}{H}^{\wedge}} = 0 \quad .$$

Thus there exists $T(\epsilon)$ such that if $u_{\epsilon}^{\wedge}(\underline{x}) = u_{T(\epsilon)}^{\wedge}(\underline{x})$ then (2) is satisfied. We have

$$(3) \qquad \|[U^{\wedge}(t,j) - U^{\wedge}(j)]u_{\epsilon}^{\wedge}\|_{\underset{\sim}{H}^{\wedge}}^2 = (2\pi)^{-n} \int_{T_{\epsilon}^{-1} \leq |\underline{x}| \leq T_{\epsilon}} |G(t,j,\underline{x}) - G(j,\underline{x})|^2 |u_{\epsilon}^{\wedge}(\underline{x})|^2 d\underline{x}.$$

Since $|G(t,j,\underline{x}) - G(j,x)|^2$ goes to zero as t goes to infinity boundedly for $T_{\epsilon}^{-1} \leq |\underline{x}| < T_{\epsilon}$ we can apply the Lebesgue limit theorem once again to show that the right hand side of (3) goes to zero as t goes to infinity.

The corresponding result for $V^{\wedge}(t,j)$ and $V^{\wedge}(j)$ is proved in the same way.

6. $\underline{N}^{\wedge}(t)$

Theorem 6a. $\underline{N}_F^{\wedge}(t) \rightrightarrows 0$ as $t \to \infty$.

Proof. Since $\underline{N}_F^{\wedge}(t) = t^{-\omega}L(t^{-1})\underline{\chi}(t)N_E^{\wedge}(t)\underline{\chi}(t)^*$ it is sufficient to show that

$$t^{-\omega}L(t^{-1})N_E^{\wedge}(t) \rightrightarrows 0 \qquad \text{as } t \to \infty \ .$$

Let $u^{\wedge}, v^{\wedge} \in \underset{\sim}{E}^{\wedge}$ with $\|u^{\wedge}\|_{\underset{\sim}{E}^{\wedge}} = \|v^{\wedge}\|_{\underset{\sim}{E}^{\wedge}} = 1$. We have

$$\langle N_{E}^{\wedge}(t)u^{\wedge} | v^{\wedge}\rangle_{\underset{\sim}{H}^{\wedge}} = (2\pi)^{-n} \int_{\underline{T}_n} h(\underline{\theta})[E^{\wedge}(t)u^{\wedge}\cdot(\underline{\theta})]\overline{[E^{\wedge}(t)v^{\wedge}(\underline{\theta})]}d\underline{\theta} \quad .$$

Given $\varepsilon > 0$ let

$$B(t,\varepsilon) = \{\underline{\theta} \in \underline{T}_n : t^{-\omega}L(t^{-1})|h(\underline{\theta})| > \varepsilon\} \quad ,$$

$$C(t,\varepsilon) = \{\underline{\theta} \in \underline{T}_n : t^{-\omega}L(t^{-1})|h(\underline{\theta})| \leq \varepsilon\} \quad .$$

Then

$$|\langle N_{E}^{\wedge}(t)u|v\rangle_{\underset{\sim}{E}^{\wedge}}| \leq I_1 + I_2$$

where

$$I_1 = (2\pi)^{-n} \int_{B(t,\varepsilon)} |h(\underline{\theta})\{E^{\wedge}(t)u^{\wedge}\cdot(\underline{\theta})\}\overline{\{E^{\wedge}(t)v^{\wedge}\cdot(\underline{\theta})\}}| d\underline{\theta} \quad ,$$

$$I_2 = (2\pi)^{-n} \int_{C(t,\varepsilon)} |h(\underline{\theta})\{E^{\wedge}(t)u^{\wedge}\cdot(\underline{\theta})\}\overline{\{E^{\wedge}(t)v^{\wedge}\cdot(\underline{\theta})\}}| d\underline{\theta} \quad .$$

We have

$$t^{-\omega}L(t^{-1})I_2 \leq \varepsilon\|E^{\wedge}(t)u^{\wedge}\|_{\underset{\sim}{E}^{\wedge}}\|E^{\wedge}(t)v^{\wedge}\|_{\underset{\sim}{E}^{\wedge}} \quad ,$$

and thus,

$$t^{-\omega}L(t^{-1})I_2 \leq \varepsilon\|W\|_{\infty}\|u^{\wedge}\|_{\underset{\sim}{E}^{\wedge}}\|W\|_{\infty}\|v^{\wedge}\|_{\underset{\sim}{E}^{\wedge}} = \varepsilon\|W\|_{\infty}^{2} \quad ,$$

where we have used (2) of §1. Similarly if $t \geq 1$

$$I_1 \leq (2\pi)^{-n} \int_{B(t,\epsilon)} |h(\underline{\theta})|\, d\underline{\theta}\} \|E^{\wedge}(t)u\|_\infty \|E^{\wedge}(t)v\|_\infty \quad ,$$

$$\leq t^n \|W^{\#}\|_2^2 \,(2\pi)^{-n} \int_{B(t,\epsilon)} |h(\underline{\theta})|\, d\underline{\theta} \quad ,$$

see (3) of §1. Let $\tau = t^{\omega} L(t^{-1})^{-1}\epsilon$. By assumption (8) of §1

$$\int_{B(t,\epsilon)} |h(\underline{\theta})|\, d\underline{\theta} = o\{L(\tau^{-1/\omega})^{-n/\omega}\, \tau^{(\omega-n)/\omega}\} \quad ,$$

$$= o\{L[t^{-1}L(t^{-1})^{1/\omega}\, \epsilon^{-1/\omega}]\}^{-n/\omega}\{t^{\omega}L(t^{-1})^{-1}\epsilon\}^{(\omega-n)/\omega} \ .$$

Since $L(t)$ is slowly oscillating at 0 given any $\delta > 0$ the inequality

$$|\log L(t^{-1})| \leq \delta |\log t^{-1}|$$

holds for all sufficiently large t. Since δ is arbitrary this implies that

$$\log[L(t^{-1})^{1/\omega}\epsilon^{-1/\omega}] = o[\log t^{-1}]$$

Using (5) and (6) of §1 we see that

$$L[t^{-1}L(t^{-1})^{1/\omega}\epsilon^{-1/\omega}] \sim L(t^{-1}) \qquad \text{as } t \to \infty \ .$$

It follows that

$$t^{-\omega}L(t^{-1})I_1 = o[t^{-\omega}L(t^{-1})t^{n}L(t^{-1})^{-n/\omega}t^{\omega-n}L(t^{-1})^{-(\omega-n)/\omega}]$$

(2)

$$= o(1) \qquad\qquad \text{as} \quad t \to \infty \ .$$

Together (1) and (2) imply that

$$\lim_{t \to +\infty} \|\underline{N}_{\underline{F}}^{\wedge}(t)\| \le \epsilon \|W\|_{\infty}^{2} \quad ;$$

but $\epsilon > 0$ is arbitrary, etc.

7. The Asymptotic Formula

Let

$$\underline{S}_{\underline{F}}^{\wedge}(t) = \int \lambda \, dC(t;\lambda) \quad ,$$

$$\underline{S}_{\underline{F}}^{\wedge} = \int \lambda \, dC(\lambda) \quad ,$$

be the spectral resolutions of the bounded self-adjoint operators $\underline{S}_{\underline{F}}^{\wedge}(t)$ and $\underline{S}_{\underline{F}}^{\wedge}$, where $\underline{S}_{\underline{F}}^{\wedge}$ and $\underline{S}_{\underline{F}}^{\wedge}(t)$ are defined in §2. We assume that $C(t;\lambda)$ and $C(\lambda)$ are right continuous and that $C(t;\infty) = C(\infty) = I$.

Theorem 7a. Under the assumptions of §1

$$C(t,\lambda) \to C(\lambda) \qquad \text{as } t \to +\infty$$

for all $\lambda \notin \sigma_{p}(\underline{S}_{\underline{F}}^{\wedge})$.

Proof. This follows from the results of §1-§5 and Theorem 1b of V.

Lemma 7b. Let $\underline{u}^{\wedge}(t) \in \underline{\underline{H}}^{\wedge}$, $\|\underline{u}^{\wedge}(t)\|_{\underline{\underline{H}}^{\wedge}} = 1$ for $1 \le t < \infty$.
Fix j,k $1 \le j, k \le p$, and set

$$w_{j,k}^{\wedge}(t) = F^{\wedge}(t,j,k) u_{k}^{\wedge}(t) \quad .$$

Note that $w^{\hat{}}_{j,k}(t) = w^{\hat{}}_{j,k}(t;\underline{x})$ is a function of $\underline{x}$. Then, given $\rho > 0$, there exists $T_\rho \geq 1$ such that the functions $\{w^{\hat{}}_{j,k}(t;\underline{x})\}_{t > T_\rho}$ are equicontinuous for $\underline{x} \in S_\rho$, the sphere with center at the origin and radius ρ.

Proof. Given $\epsilon > 0$ choose $r > 0$ so large that

$$\int_{|\underline{y}| \geq r-1} W^{\#}(\underline{y})^2 d\underline{y} \leq \epsilon^2 .$$

If t is sufficiently large, $t \geq T_\rho$, then $R^t_j \supset S_\rho$. For $t \geq T_\rho$ and $\underline{x} \in S_\rho$ we have, see Lemma 3a,

$$w^{\hat{}}_{j,k}(t;\underline{x}) = I_1(t,\underline{x}) + I_2(t,\underline{x})$$

where

$$I_1(t,\underline{x}) = t^{-n} \sum_{|\underline{m}| \leq tr} e^{it^{-1}\underline{m}\cdot[\underline{x}+t(\underline{\theta}_j-\underline{\theta}_k)]} W(t^{-1}\underline{m}) u_k(t,\underline{m}) ,$$

$$I_2(t,\underline{x}) = t^{-n} \sum_{|\underline{m}| > tr} e^{it^{-1}\underline{m}\cdot[\underline{x}+t(\underline{\theta}_j-\underline{\theta}_k)]} W(t^{-1}\underline{m}) u_k(t,\underline{m}) .$$

Here

$$u_k(t,\underline{m}) = (2\pi)^{-n} \int_{R^t_k} u^{\hat{}}_k(t,\underline{x}')e^{-it^{-1}\underline{m}\cdot\underline{x}'} d\underline{x}' .$$

It is easy to verify that

$$\sum_{\underline{m}} |u_k(t,\underline{m})|^2 = t^n(2\pi)^{-n} \int_{R^t_k} |u^{\hat{}}_k(t,\underline{x}')|^2 d\underline{x}' \leq t^n .$$

It follows that if $t \geq 1$

$$\left|\frac{\partial}{\partial x_s}\right| |I_1(t,\underline{x})| \leq t^{-n} \sum_{|\underline{m}| \leq tr} t^{-1}|m_s| \, |W(t^{-1}\underline{m})| \, |u_k(t,\underline{m})| \quad ,$$

$$\leq rt^{-n}\Big\{\sum_{\underline{m}} |W(t^{-1}\underline{m})|^2\Big\}^{\frac{1}{2}} \Big\{\sum_{\underline{m}} |u_k(t,\underline{m})|^2\Big\}^{\frac{1}{2}} \quad ,$$

$$\leq rt^{-n}\Big\{t^{n/2}\|W^{\#}\|_2\Big\} \Big\{t^{n/2}\Big\} = r\|W^{\#}\|_2 \quad .$$

Again

$$|I_2(t,\underline{x})| \leq t^{-n} \sum_{|\underline{m}| > tr} |W(t^{-1}\underline{m})| \, |u_k(t,\underline{m})| \quad ,$$

$$\leq t^{-n}\Big\{\sum_{|\underline{m}| > tr} |W(t^{-1}\underline{m})|^2\Big\}^{\frac{1}{2}} \Big\{\sum_{\underline{m}} |u_k(t,\underline{m})|^2\Big\}^{\frac{1}{2}} \quad .$$

Since $t \geq 1$

$$\sum_{|\underline{m}| > tr} |W(t^{-1}\underline{m})|^2 \leq t^n \int_{|\underline{y}| \geq r-1} W^{\#}(\underline{y})^2 d\underline{y} \leq t^n \epsilon^2 \quad ,$$

from which we obtain

$$|I_2(t,\underline{x})| \leq t^{-n}\{t^n \epsilon^2\}^{\frac{1}{2}} \{t^n\}^{\frac{1}{2}} \leq \epsilon \quad .$$

Our assertion is a consequence of these inequalities.

Corollary 7c. If $j \neq k$ and if $\underset{\sim}{P}_1$ is a subsequence of $\underset{\sim}{P}$, then $\underset{\sim}{P}_1$ contains a subsequence $\underset{\sim}{P}_2$ such that

$$\lim_{\underset{\sim}{P}_2} w^{\wedge}_{j,k}(t;\underline{x}) = 0$$

uniformly in every sphere S_ρ .

Proof. By Lemma 7b and Arzela's theorem $\underset{\sim}{P}_1$ contains a subsequence $\underset{\sim}{P}_2$ such that as $t \to \infty$ in $\underset{\sim}{P}_2$ $w^{\wedge}_{j,k}(t;\underline{x})$ converges to a function $w^{\wedge}_{j,k}(\underline{x})$ uniformly in every sphere. By Fatou's lemma $w^{\wedge}_{j,k} \in \underset{\sim}{H}^{\wedge}$. If $w^{\wedge}_{j,k}(x) \neq 0$ there exists $v^{\wedge}_j \in \underset{\sim}{H}^{\wedge}$ such that $\langle v^{\wedge}_j | w^{\wedge}_{j,k} \rangle_{\underset{\sim}{H}^{\wedge}} \neq 0$. However, we have

$$\begin{aligned} \langle v^{\wedge}_j | w^{\wedge}_{j,k} \rangle_{\underset{\sim}{H}^{\wedge}} &= \lim_{\underset{\sim}{P}_2} \langle v^{\wedge}_j | F^{\wedge}(t,j,k) u^{\wedge}_k(t) \rangle_{\underset{\sim}{H}^{\wedge}} \quad , \\ &= \lim_{\underset{\sim}{P}_2} \langle F^{\wedge}(t,j,k)^* v^{\wedge}_j | u^{\wedge}_k(t) \rangle_{\underset{\sim}{H}^{\wedge}} \quad , \\ &= 0 \ , \end{aligned}$$

since by Theorem 3c $\|F^{\wedge}(t,j,k)^* v^{\wedge}_j\|_{\underset{\sim}{H}^{\wedge}} \longrightarrow 0$ as $t \to \infty$. This contradiction proves our assertion.

Corollary 7d. Let $j \neq k$ and/or $j \neq m$, and let

$$J(t) = \langle U^{\wedge}(t,j) F^{\wedge}(t,j,k) u^{\wedge}_k(t) | V^{\wedge}(t,j) F^{\wedge}(t,j,m) u^{\wedge}_m(t) \rangle_{\underset{\sim}{H}^{\wedge}} \ .$$

Then

$$\lim_{t \to \infty} J(t) = 0 \ .$$

Proof. Given $\epsilon > 0$ choose ρ so large that $G(x) \leq \epsilon$ if $|\underline{x}| \geq \rho$. Then

$$J(t) = J_1(t) + J_2(t)$$

where

$$J_1(t) = (2\pi)^{-n} \int_{|x| \leq \rho} G(t,j,\underline{x}) w^{\hat{}}_{j,k}(t;\underline{x}) \overline{w^{\hat{}}_{j,m}(t;\underline{x})} d\underline{x} \quad ,$$

$$J_2(t) = (2\pi)^{-n} \int_{|\underline{x}| > \rho} G(t,j,\underline{x}) w^{\hat{}}_{j,k}(t;\underline{x}) \overline{w^{\hat{}}_{j,m}(t;\underline{x})} d\underline{x} \quad .$$

By Corollary 7c $J_1(t) \to 0$ as $t \to \infty$. On the other hand making use of Theorem 3b and (3) of §4 we see that

$$|J_2(t)| \leq \epsilon \|w^{\hat{}}_{j,k}(t)\|_{\underset{\sim}{H}^{\hat{}}} \|w^{\hat{}}_{j,m}(t)\|_{\underset{\sim}{H}^{\hat{}}} \leq \epsilon \|W\|_\infty^2 \quad .$$

Thus

$$\overline{\lim_{t \to +\infty}} |J_2(t)| \leq \epsilon \|W\|_\infty^2 \quad ,$$

etc.

Theorem 73. Let $\underline{u}^{\hat{}}(t) \in \underset{\approx}{H}^{\hat{}}$, $\|\underline{u}^{\hat{}}(t)\|_{\underset{\approx}{H}^{\hat{}}} = 1$ for $1 \leq t \infty$ and let $\underset{\sim}{P}_1$ be a subsequence of $\underset{\sim}{P}$. If

$$\langle \underline{S}_{\underline{F}}(t) \underline{u}^{\hat{}}(t) | \underline{u}^{\hat{}}(t) \rangle \geq m_1 > 0 \qquad t \in \underset{\sim}{P}_1 \quad ,$$

or if

$$\langle \underline{S}_{\underline{F}}(t)\underline{u}^{\wedge}(t)|\underline{u}^{\wedge}(t)\rangle \leq -m_1 < 0 \qquad t \in P_1 \quad ,$$

and if

$$\underline{u}^{\wedge}(t) \longrightarrow \underline{u}^{\wedge} \qquad \text{as} \quad t \to \infty \quad \text{in} \quad P_1 \; ,$$

then $\underline{u}^{\wedge} \not\equiv 0$.

Proof. We need only consider the first case. We have

$$\langle \underline{S}_{\underline{F}}^{\wedge}(t)\underline{u}^{\wedge}(t)|\underline{u}^{\wedge}(t)\rangle_{H^{\wedge}} = \sum_{j,k,m=1}^{p} \langle U^{\wedge}(t,j)\underline{F}^{\wedge}(t,j,k)u_k^{\wedge}(t)|V^{\wedge}(t,j)\underline{F}^{\wedge}(t,j,m)u_m^{\wedge}(t)\rangle_{H^{\wedge}} \; .$$

Using Corollary 7d we see that if as before $w_{j,j}^{\wedge}(t) = \underline{F}^{\wedge}(t,j,j)u_j^{\wedge}(t)$ then

$$\underline{\lim}_{P_1} \sum_{j=1} (2\pi)^{-n} \int_{R_n} G(t,j,\underline{x})|w_{j,j}^{\wedge}(t,\underline{x})|^2 d\underline{x} \geq m_1 \quad ,$$

and à fortiori

$$(1) \qquad \underline{\lim}_{P_1} \sum_{j=1}^{q} (2\pi)^{-n} \int_{R_n} G(t,j,\underline{x})|w_{j,j}^{\wedge}(t,\underline{x})|^2 d\underline{x} \geq m_1 \quad .$$

By Lemma 7b and Arzela's theorem there is a subsequence P_2 of P_1 such that for some continuous function $w_{j,j}^{\wedge}(\underline{x})$ we have

$$w_{j,j}^{\wedge}(t;\underline{x}) \longrightarrow w_{j,j}^{\wedge}(\underline{x}) \qquad \text{as} \quad t \to \infty \quad \text{in} \quad P_2$$

uniformly on every compact set in $\underline{R}_n$ for $1 \leq j \leq p$. Since $w^{\wedge}_{j,j}(t) = F^{\wedge}(j,j)u^{\wedge}_j \rightarrow F^{\wedge}u^{\wedge}_j$, $\underline{u}^{\wedge} \equiv 0$ would imply that

(2) $$w^{\wedge}_{j,j}(t;\underline{x}) \rightarrow 0 \qquad \text{as} \quad t \to \infty \quad \text{in} \quad \underset{\sim}{P}_2$$

uniformly on compact sets in $\underline{R}_n$ for $j = 1,\ldots,q$. Given $\epsilon > 0$ choose ρ so that

(3) $$G(\underline{x}) < \epsilon \qquad \text{if} \quad |\underline{x}| \geq \rho \quad .$$

Split each integral in (1) into two pieces corresponding to the ranges $|x| < \rho$ and $|\underline{x}| > \rho$. Using (2) when the range is $|\underline{x}| \leq \rho$ and (3) and Theorem 3b when the range is $|\underline{x}| \geq \rho$ we find that

(4) $$\underset{\underset{\sim}{P}_2}{\underline{\lim}} \sum_{j=1}^{q} (2\pi)^{-n} \int_{R_n} G(t,j,\underline{x}) |w^{\wedge}_{j,j}(t;\underline{x})|^2 d\underline{x} \leq \epsilon \|W\|_{\infty}^2 \ .$$

Since $\epsilon > 0$ is arbitrary (1) and (4) are in contradiction, and our assertion is proved.

Theorem 7f. Under the assumptions of §1 $\dim[I-C(\lambda)] < \infty$ if $\lambda > 0$, $\dim C(\lambda) < \infty$ if $\lambda < 0$, and

$$\lim_{t \to \infty} \dim[I-C(t,\lambda)] = \dim[I-C(\lambda)] \qquad \lambda > 0,\ \lambda \in \sigma_p(\underline{S}^{\wedge}_{\underline{F}}) \ ,$$

$$\lim_{t \to \infty} \dim C(t,\lambda) = \dim C(\lambda) \qquad \lambda < 0,\ \lambda \notin \sigma_p(\underline{S}^{\wedge}_{\underline{F}}) \ .$$

Proof. This follows from Theorem 7a, Theorem 7c, and Theorem 2a of V.

We recall that

$$\underline{T}_{\underline{F}}^{\wedge}(t) = \underline{S}_{\underline{F}}^{\wedge}(t) + \underline{N}_{\underline{F}}^{\wedge}(t) \quad .$$

Let

$$\underline{T}_{\underline{F}}^{\wedge}(t) = \int \lambda dC'(t,\lambda)$$

be the spectral resolution of $\underline{T}_{\underline{F}}^{\wedge}(t)$ normalized in the usual fashion.

Theorem 7g. Under the assumptions of §1

$$C'(t,\lambda) \rightarrow C(\lambda) \qquad \text{as} \quad t \to \infty$$

for all $\lambda \notin \sigma_p(\underline{S}_{\underline{F}}^{\wedge})$, and

$$\lim_{t \to \infty} \dim[I-C'(t,\lambda)] = \dim[I-C(\lambda)] \qquad \lambda > 0, \lambda \notin \sigma_p(\underline{S}_{\underline{F}}^{\wedge}) ,$$

$$\lim_{t \to \infty} \dim C'(t,\lambda) = \dim C(\lambda) \qquad \lambda < 0, \lambda \notin \sigma_p(\underline{S}_{\underline{F}}^{\wedge}) .$$

Proof. This follows from Theorem 6a, Theorem 7f and Corollary 2c of VI.

What remains is to introduce the notation necessary to obtain from Theorem 7g a statement about the eigen values of $\underline{T}_{\underline{E}}^{\wedge}(t)$. It is clear that

$$\underline{S}_{\underline{F}}^{\wedge} = S_F^{\wedge}(1) \oplus \cdots \oplus S_F^{\wedge}(p)$$

where $S_F^{\wedge}(k)$ is completely determined up to sign by ω, $\Phi_k(\underline{x})$ and $W(\underline{y})$.

The sign depends on whether $1 \leq k \leq q$, or $q < k \leq p$. By Theorem 7f the operators $S_F^{\hat{}}(k)$, $1 \leq k \leq q$ $(q < k \leq p)$ are positive (negative), completely continuous, and self-adjoint on $\underset{\sim}{H}^{\hat{}}$. For $1 \leq k \leq q$ let

$$\nu^+(k,1) > \nu^+(k,2) \geq \cdots , \quad \lim_{j \to \infty} \nu^+(k,j) = 0$$

be the eigen values of $S_F^{\hat{}}(k)$, repreated according to their multiplicities. Let $\mu^+(k)$, $k = 1,2,\ldots$, be the non-increasing rearrangement of $\nu^+(k,j)$ where $1 \leq k \leq q$, $j = 1,2,\ldots$. Similarly for $q < k \leq p$ let

$$\nu^-(k,1) \leq \nu^-(k,2) < \cdots , \quad \lim_{j \to \infty} \nu^-(k,j) = 0$$

be the eigen values of $S_F^{\hat{}}(k)$ repeated according to their multiplicities, and let $\mu^-(k)$, $k = 1,2,\ldots$, be the non-decreasing rearrangement of $\nu^-(k,j)$ where $q < k < p$, $j = 1,2,\ldots$.

Let $\lambda^+(t,1) \geq \lambda^+(t,2) \geq \cdots$ $(\lambda^-(t,1) \leq \lambda^-(t,2) \leq \cdots)$ be the positive (negative) eigen values of $\underline{T}_{\underline{E}}^{\hat{}}(t)$ repeated according to their multiplicities.

Corollary 7h. We have

$$\left.\begin{aligned} \lambda^+(t,k) &\sim t^{\omega} L(t^{-1}) \mu^+(k) \\ \lambda^-(t,k) &\sim t^{\omega} L(t^{-1}) \mu^-(k) \end{aligned}\right\} \quad \text{as } t \to \infty$$

for each $k = 1,2,\ldots$.

Note that if $q = 0$ $(q = p)$ no statement is made about the positive (negative) eigen values of $\underline{T}_{\underline{E}}^{\hat{}}(t)$.

8. The Fourier transform

This section is devoted to the Fourier transform case, which is much simpler than the Fourier series case we have dealt with in §1 - §7. We confine ourselves to formulating the problem and stating its solution, leaving the details of proof to the reader.

We make use of Hilbert spaces $\underset{\sim}{L}$, $\underset{\sim}{L}^{\wedge}$, $\underset{\approx}{H}^{\wedge}$ and $\underset{\approx}{H}$, and the mappings between them defined in §1 of III.

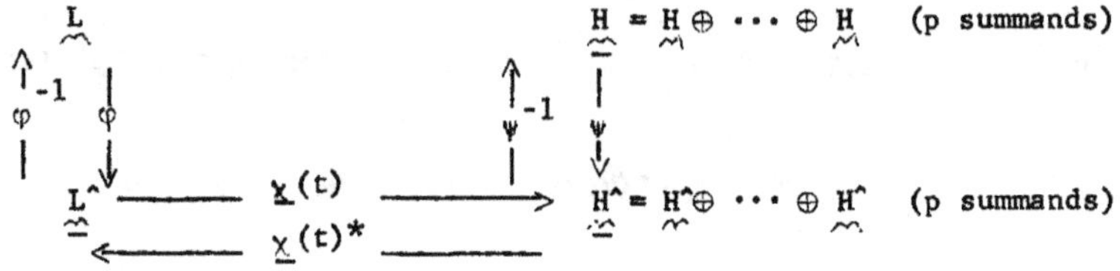

Let $W(\underline{y})$ be a complex function on $\underline{R}_n$ belonging to $L^{\infty}(\underline{R}_n) \cap L^2(\underline{R}_n)$. We define an operator $E(t)$ on $\underset{\sim}{L}$ by

$$E(t)\, u \cdot (\underline{\eta}) = W(t^{-1}\underline{\eta})u(\underline{\eta}) \quad .$$

It is apparent that if $0 < t < \infty$

$$\|E(t)u\|_{\underset{\sim}{L}} \leq \|W\|_{\infty} \|u\|_{\underset{\sim}{L}} \quad ,$$

$$\|E(t)u\|_{1} \leq t^{n/2} \|W\|_{2} \|u\|_{\underset{\sim}{L}} \quad .$$

We define $E^{\wedge}(t)$ by the formula

$$E^{\wedge}(t)u^{\wedge} \cdot (\underline{\xi}) = [\varphi\, E(t)\varphi^{-1}u^{\wedge}] \cdot (\underline{\xi}) \qquad u^{\wedge} \in \underset{\sim}{L}^{\wedge} \; .$$

Clearly

$$\|E^{\wedge}(t)u^{\wedge}\|_{\underset{\sim}{L}^{\wedge}} \leq \|W\|_{\infty} \|u^{\wedge}\|_{\underset{\sim}{L}^{\wedge}} \quad ,$$

(1)

$$\|E^{\wedge}(t)u^{\wedge}\|_{\infty} \leq t^{n/2}\|W\|_{2} \|u^{\wedge}\|_{\underset{\sim}{L}^{\wedge}} \quad .$$

For $f(\underline{\xi})$ any function in $L^1(\underline{R}_n) + L^\infty(\underline{R}_n)$ consider the sesquilinear form

$$(u^\wedge,v^\wedge)_t = (2\pi)^{-n} \int_{\underline{R}_n} f(\underline{\xi})[E^\wedge(t)u^\wedge\cdot(\underline{\xi})]\overline{[E^\wedge(t)v^\wedge\cdot(\underline{\xi})]}\,d\underline{\xi} \quad .$$

It follows from (1) that $(u^\wedge,v^\wedge)_t$ is bounded for each $t > 0$. Since $(u^\wedge,v^\wedge)_t = (v^\wedge,u^\wedge)_t$ there exists a bounded self-adjoint operator $T_E^\wedge(t)$ on $\underset{\sim}{L}^\wedge$ such that

$$\langle T_E^\wedge(t)u^\wedge | v^\wedge\rangle_{\underset{\sim}{L}^\wedge} = (u^\wedge,v^\wedge)_t \quad .$$

As before our goal is to describe how the largest positive and negative eigen values of $T_E^\wedge(t)$ behave as $t \to +\infty$, which is possible by our methods only when $f(\underline{\xi})$ has the structure we now describe.

Let $L(t)$ be as in §1 of VI, and let us fix ω, $0 < \omega < n$.

Definition 1a. A real function $h(\underline{\xi})$ on $\underline{R}_n$ will be said to be "negligible" if

$$\int_{A(\tau)} |h(\underline{\xi})|\,d\underline{\xi} = o\{L(\tau^{-1/\omega})^{-n/\omega}\tau^{(\omega-n)/\omega}\} \quad \text{as} \quad \tau \to \infty$$

where

$$A(\tau) = \{\underline{\xi} \in \underline{R}_n : |h(\underline{\xi})| > \tau\} .$$

If $h(\underline{\xi})$ is negligible then $h \in L^\infty(\underline{R}_n) + L^1(\underline{R}_n)$.

We fix two integers $0 \leq q \leq p$ and p distinct points $\{\underline{\xi}_k\}_1^p$ in $\underline{R}_n$. Let

$$\underline{R}_n = \bigcup_{k=1}^{p} R_k$$

be a representation of $\underline{R}_n$ as a union of p non-overlapping, possibly infinite, rectangles with sides parallel to the coordinate axis such that $\underline{\xi}_k$ lies in the interior of R_k for $1 \leq k \leq p$. Let $\Phi_k(\underline{\xi})$ and $\gamma(k)$ be as in §1 of VII. We define a function $g(\underline{\xi})$ on $\underline{R}_n$ by setting

$$g(\underline{\xi}) = \gamma(k) g_k(\underline{\xi}) \qquad \underline{\xi} \in R_k \quad k = 1,\dots,p$$

where

$$g_k(\underline{\xi}) = |\underline{\xi} - \underline{\xi}_k|^{-\omega} L(|\underline{\xi} - \underline{\xi}_k|)^{-1} \Phi_k(\underline{\xi} - \underline{\xi}_k) .$$

Our assumption on $f(\underline{\xi})$ is that

$$f(\underline{\xi}) = g(\underline{\xi}) + h(\underline{\xi})$$

where $g(\underline{\xi})$ and $h(\underline{\xi})$ are as described above.

By arguments like these given in §1 - §7 we obtain the following.

Theorem 1b. Under these assumptions we have

$$\left.\begin{aligned} \lambda^+(t,k) &\sim t^{\omega} L(t^{-1}) \mu^+(k) \\ \lambda^-(t,k) &\sim t^{\omega} L(t^{-1}) \mu^-(k) \end{aligned}\right\} \quad \text{as} \quad t \to \infty$$

for each $k = 1,2,\dots$.

Here $\lambda^+(t,1) \geq \lambda^+(t,2) \geq \cdots$ are the largest positive eigen values of $T_{\hat{E}}(t)$ and $\lambda^-(t,1) \leq \lambda^-(t,2) \leq \cdots$ are the smallest negative eigen values of $T_{\hat{E}}(t)$. $\mu^+(k)$ and $\mu^-(k)$ are defined exactly as in §7.

9. Asymptotic formulas for large eigen value problem continued

The solution of the "large eigen value" problem given in the preceding sections involves the eigen values of the special integral operators

$$T_k = M(W)\varphi^{-1}E_k\varphi M(\bar{W}) \qquad k = 1,\ldots,p$$

where φ is the Fourier transform on $\underline{R}_n$ n-dimensional Euclidean space and where

$$M(W)u(\underline{x}) = W(\underline{x})u(\underline{x}) ,$$

$$E_k\hat{u}(\underline{y}) = |\underline{y}|^{-\omega}\Phi_k(\underline{y}) .$$

Here $0 < \omega < n$, and the $\Phi_k(\underline{y})$'s are the bounded, non-negative, measurable function, homogeneous of degree 0, defined in §1 of Chapter VI. We recall that

$$\nu^+(k,1) \geq \nu^+(k,2) \geq \cdots > 0$$

are the positive eigen values of T_k if $1 \leq k \leq q$, while

$$-\nu^-(k,1) \geq -\nu^-(k,2) \geq \cdots > 0$$

are the positive eigen values of T_k if $q < k \leq p$. What is of interest is that (under one additional restriction on W) there exist quite explicit asymptotic formulas for the $\nu^+(k,j)$'s and $\nu^-(k,j)$'s as $j \to \infty$. These formulas are discussed in § 7 of [10]. The exposition there is incomplete in that various assertions are made which are not there verified. However, the correctness of these assertions can be checked quite routinely. Let $N^+[\epsilon,T_k]$ be the number of eigen values of T_k greater than $\epsilon > 0$, and let

$$\Psi_k(\varepsilon) = \left|\{\underline{x},\underline{y}) : |W(\underline{x})|^2 |\underline{y}|^{-\omega}\Phi_k(\underline{y}) > \varepsilon\}\right|_{\underline{R}_n \times \underline{R}_n} ,$$

where $|\{\cdot\}|_{\underline{R}_n \times \underline{R}_n}$ is the Lebesgue measure of the set $\{\cdot\}$ in $\underline{R}_n \times \underline{R}_n$.

Theorem 9a. If $W(x) \in L^2(\underline{R}_n) \cap L^\infty(\underline{R}_n)$ has bounded support in $\underline{R}_n$ then for any fixed $\delta > 0$

$$N^+[\varepsilon, T_k] \lesssim \Psi_k\left(\frac{\varepsilon}{1 - \delta}\right)$$

$$\text{as} \quad \varepsilon \to 0+ ,$$

$$N^+[\varepsilon, T_k] \gtrsim \Psi_k\left(\frac{\varepsilon}{1 + \delta}\right)$$

where $a(\varepsilon) \gtrsim b(\varepsilon)$ as $\varepsilon \to 0+$ means that $\overline{\lim}_{\varepsilon \to 0+} a(\varepsilon) b(\varepsilon)^{-1} \geq 1$, etc.

One can also show that the formulas above are valid if $W(x)$ is the characteristic function of a set of finite positive measure which need not have bounded support.

Remark. In [10] it is assumed that $W(x)$ is non-negative, while here we have assumed only that $W(x)$ is complex valued. This is inconsequential since, if

$$U_k = M(|W|) \circ \omega^{-1} E_k \circ \omega M(|W|) ,$$

an evident trivial argument shows that T_k and U_k are unitarily equivalent.

Bibliography

1. Baxley, John "Extreme eigenvalues of Toeplitz matrices associated with Laguerre polynomials," Arch. Rat. Mech. Anal., 30 (1968), pp. 308-320.

2. "Extreme eigenvalues of Toeplitz matrices associated with certain orthogonal polynomials," SIAM J. Math. Anal., 2 (1971), pp. 470-482.

3. Davis, Jeffrey "Extreme eigenvalues of Toeplitz operators of the Hankel type, I," Jour. Math. and Mech., 14 (1965), pp. 245-276.

4. "Extreme eigenvalues of Toeplitz operators of the Hankel type, II," Trans. Amer. Math. Soc., 116 (1965), pp. 267-299.

5. Grenander, U. and G. Szegö. Toeplitz Forms and their Applications. University of California Press, 1958.

6. Hirschman, I. I. "Extreme eigenvalues of Toeplitz forms associated with ultraspherical polynomials," J. of Math. and Mech., 13 (1964), pp. 249-282.

7. "Extreme eigenvalues of Toeplitz forms associated with Jacobi polynomials," Pacific J. Math., 14 (1964), pp. 107-161.

8. "Extreme eigenvalues of Toeplitz forms associated with orthogonal polynomials," Jour. d'Analyse Math., 12 (1964), pp. 187-242.

9. "Integral equations on certain compact symmetric spaces," SIAM Journal Math. Anal., 3 (1972), pp. 314-343.

10. "On the eigenvalues of certain integral operators," SIAM Journal Math. Anal., 6 (1975), pp. 1024-1050.

11. Kac, Marc "On some connections between probability theory and integral equations," Proc. Second Berkeley Symp. on Math. Statistics and Probability, Univ. of California Press, Berkeley, 1951, pp. 189-215.

12. "Distributions of eigenvalues of certain integral operators," Michigan Math. J., 3 (1955-56), pp. 141-148.

13. Kac, M., Murdock, W.L., and Szegö, G. "On the eigenvalues of certain Hermitian forms," J. Rat. Mech. Anal., 2 (1953), pp. 767-800.

14. Katznelson, Y. An Introduction to Harmonic Analysis. John Wiley and Sons, Inc., New York, 1968.

15. Kesten, H. "On the extreme eigenvalues of translation kernels and Toeplitz matrices," J. d'Analyse Math., 10 (1962-63), pp. 117-138.

16. Krieger, H. A. "Toeplitz operators on locally compact Abelian groups," J. Math. and Mech., 14 (1965), pp. 439-478.

17. Liang, D. Thesis at Washington University, "Eigenvalue Distributions of Toeplitz Operators", 1974.

18. Parter, S. V. "On the extreme eigenvalues of truncated Toeplitz matrices, Bull. Amer. Math. Soc., 67 (1961), pp. 191-196.

19. "On the extreme eigenvalues of Toeplitz matrices," Trans. Amer. Math. Soc., 100 (1961), pp. 263-276.

20. "Extreme eigenvalues of Toeplitz forms and applications to elliptic differential equations," Trans. Amer. Math. Soc., 99 (1961) pp. 153-193.

21. "On the eigenvalues of certain generalizations of Toeplitz matrices," Arch. Rat. Mech. and Anal., 11 (1962), pp. 244-257.

22. "Remarks on the extreme eigenvalues of Toeplitz forms associated with orthogonal polynomials," _Jour. Math. Anal. and Appl._, 12 (1965), pp. 456-470.

23. Riesz, F. and B. Sz.-Nagy. Functional Analysis. Frederick Ungar Publishing Co., New York, 1955.

24. Rosenblatt, M. "Asymptotic behaviour of eigenvalues for a class of integral equations with translation kernels," Proc. Symposium on Time Series Analysis, 1963, J. Wiley and Sons, New York.

25. "Some results on the asymptotic behaviour of eigenvalues of a class of integral equations with translation kernels," _J. Math. Mech._, 12 (1963), pp. 619-628.

26. Widom, H. "On the eigenvalues of certain Hermitian operators," _Trans. Amer. Math. Soc._, 88 (1958), pp. 491-522.

27. "Stable processes and integral equations," _Trans. Amer. Math. Soc._, 98 (1961), pp. 430-439.

28. "Extreme eigenvalues of translation kernels," _Trans. Amer. Math. Soc._, 100 (1961), pp. 252-262.

29. "Extreme eigenvalues of N-dimensional convolution operators," _Trans. Amer. Math. Soc._, 106 (1963), pp. 391-414.

30. "Asymptotic behaviour of the eigenvalues of certain integral equations," _Trans. Amer. Math. Soc._, 109 (1963), pp. 278-295.

31. Wilf, H. "Finite Sections of Some Classical Inequalities," _Ergebnisse der Mathematik und ihrer Grenzgebiete_, vol. 32, Springer-Verlag, 1970.

INDEX OF SYMBOLS

Because of the many special symbols used repeatedly in this paper it seems desirable to include the following index of symbols. Since the same symbol may have different (although related) meanings in successive chapters the index is broken down by chapters. The reader is advised to consult the current chapter for a given symbol and if it does not appear to read the index in reverse order to find the current meaning of the symbol.

CHAPTER I

Page 1. $\underline{T}$, $\underline{R}$, $L^2(\underline{T})$, $f^\wedge(\theta)$

2. $\underline{Z}$, $L^2(Z)$

CHAPTER II

Page:

16. $\underset{\sim}{H}$, $\langle\cdot|\cdot\rangle$, $\|\cdot\|$, $S(t)$, F, $F(t)$, $S_F(t)$, $d\psi(t,\lambda)$, S_F, $d\psi(\lambda)$

17. $\underset{\sim}{S}$, $\underset{\sim}{D}[\cdot]$, $\mathcal{S}$, $\underset{\sim}{M}$, $\longrightarrow$, $\rightharpoonup$, $\overset{\rightarrow}{\rightarrow}$

18. $(\cdot|\cdot)$, $|||\cdot|||$, w

20. P, P_1

24. $\sigma_p(\cdot)$

25. $C_o(\underline{R})$

CHAPTER III

Page:

31. Ω, $\underline{\xi} = (\xi_1,\ldots,\xi_n)$, $\underline{R}_m$, $f^\wedge(\underline{\xi})$, ξ_m, ω, $L(t)$, $\Phi_m(\underline{\xi})$

32. $\underset{\sim}{L}$, $\underline{\eta} = (\eta_1,\ldots,\eta_n)$, $\langle\cdot|\cdot\rangle_{\underset{\sim}{L}}$, $\|\cdot\|_{\underset{\sim}{L}}$

$\underset{\sim}{L}^\wedge$, $\langle\cdot|\cdot\rangle_{\underset{\sim}{L}^\wedge}$, $\|\cdot\|_{\underset{\sim}{L}^\wedge}$, Υ

33. Υ^{-1}, $\underline{\xi}\cdot\underline{\eta}$, $E(t)$, $E^\wedge(t)$, $T^\wedge$, $T^\wedge_E(t)$

34. $\lambda(j,t)$, $\underset{\sim}{H}$, $\underset{\sim}{H}^\wedge$

CHAPTER III

Page:

35. $\underline{y}$, $\langle\cdot|\cdot\rangle_{\underset{\sim}{H}}$, $\|\cdot\|_{\underset{\sim}{H}}$, $\underline{x}$, $\langle\cdot|\cdot\rangle_{\underset{\sim}{H}^\wedge}$, $\|\cdot\|_{\underset{\sim}{H}^\wedge}$,

ϕ, ϕ^{-1}, $\underset{\approx}{H}$, $\underline{u} = [u_1,\ldots,u_p]$, $\langle\cdot|\cdot\rangle_{\underset{\approx}{H}}$, $\|\cdot\|_{\underset{\approx}{H}}$, $\underset{\approx}{H}^\wedge$

36. $\underline{u}^\wedge = [u_1^\wedge,\ldots,u_p^\wedge]$, $\langle\cdot|\cdot\rangle_{\underset{\approx}{H}^\wedge}$, $\|\cdot\|_{\underset{\approx}{H}^\wedge}$,

Ψ, Ψ^{-1}, R_m, $\sigma(t,m)$, R_m^t

37. $\tau(t,m)$, $\chi(t,m)$, $\underline{\chi}(t)$

38. $\chi^*(t,m)$, $\chi^*(t)$

39. $\underline{F}^\wedge(t)$

40. $F^\wedge(t,\ell,m)$

41. $\underline{T}^\wedge(t)$

42. $f_m^t(\underline{x})$, $S^\wedge(t,m)$, $g_m^t(\underline{x})$, $\underline{S}^\wedge(t)$, $\underline{S}^\wedge$, $g_m(\underline{x})$, $S^\wedge(m)$

43. F, $\underline{F}$, $\underline{F}^\wedge$

49. $\underset{\sim}{D}(\Omega)$, $\underset{\sim}{D}^\wedge(\Omega)$, $\underset{\approx}{D}(\Omega)$, $\underset{\approx}{D}^\wedge(\Omega)$

65. $\underline{S}_{\underline{F}}^\wedge$, $d\underline{E}^\wedge(\lambda)$, $\underline{S}_{\underline{F}}^\wedge(t)$, $d\underline{E}_t^\wedge(\lambda)$

CHAPTER IV

Page:

74. Ω, $\underline{\theta} = (\theta_1,\ldots,\theta_n)$, $\underline{T}_n$, $f(\underline{\theta})$, $\underline{\theta}_m$, $\underset{\sim}{E}$, $\langle\cdot|\cdot\rangle_{\underset{\sim}{E}}$, $\|\cdot\|_{\underset{\sim}{E}}$

75. $\underline{k} = (k_1,\ldots,k_n)$, $\underset{\sim}{E}^\wedge$, $\langle\cdot|\cdot\rangle_{\underset{\sim}{E}^\wedge}$, $\|\cdot\|_{\underset{\sim}{E}^\wedge}$, φ, φ^{-1},

$\underline{k}\cdot\underline{\theta}$, Ω_t, $E(t)$, $E^\wedge(t)$, $T^\wedge$

76. $T^\wedge(t)$

77. R_m, $\sigma(t,m)$, R_m^t, $\tau t,m)$, $\chi(t,m)$, $\underline{\chi}(t)$

78. $\chi^*(t,m)$, $\underline{\chi}^*(t)$

CHAPTER IV

Page:

79. $\underline{F}^{\wedge}(t)$, $F^{\wedge}(t,\ell,m)$

80. T

81. $\underline{T}^{\wedge}(t)$, $f_m^t(\underline{x})$, $S^{\wedge}(t,m)$, $g_m^t(\underline{x})$, $\underline{S}^{\wedge}(t)$, $\underline{S}^{\wedge}$, $g_m(\underline{x})$

82. F, $\underline{F}$, $\underline{F}^{\wedge}$

93. $\underline{S}^{\wedge}_{\underline{F}}$, $dC(\lambda)$, $\underline{S}^{\wedge}_{\underline{F}}(t)$, $dC(t,\lambda)$

CHAPTER V

Page:

97. $\underset{\sim}{H}$, $F(t)$, $F(t)^*$, F, F^*, $U(t)$, U

98. $V(t)$, V, $\underline{C}$, $\underset{\sim}{H} \times \underset{\sim}{H}$, $\{\cdot,\cdot\}$, $S_F(t)$

99. S_F

CHAPTER VI

Page:

105. $\underset{\sim}{E}$, $\underset{\sim}{E}^{\wedge}$, φ, φ^{-1}, $\underline{\chi}(t)$, $\underline{\chi}(t)^*$, $w(\underline{y})$, $w^{\#}(\underline{y})$

106. $c(\underline{k},t)$, $E(t)$

107. $f^{\wedge}(\underline{\theta})$, $(\cdot,\cdot)_t$, $T^{\wedge}_E(t)$

108. ω, $L(t)$, $\underline{\theta}_k$, R_k, $\underline{\Phi}_k$, $\gamma(k)$

109. $A(\tau)$, $g^{\wedge}(\underline{\theta})$

110. $(\cdot,\cdot)'_t$, $(\cdot,\cdot)''_t$, $S^{\wedge}_E(t)$, $N^{\wedge}_E(t)$

111. $U^{\wedge}$, $V^{\wedge}$, $\underline{F}^{\wedge}(t)$, $\underline{T}^{\wedge}_{\underline{F}}(t)$, $\underline{S}^{\wedge}_{\underline{F}}(t)$, $\underline{N}^{\wedge}_{\underline{F}}(t)$,

$\underline{U}^{\wedge}(t)$, $\underline{V}^{\wedge}(t)$

112. F, $F^{\wedge}$, $\underline{F}^{\wedge}$, $G(k,\underline{x})$, $U^{\wedge}(k)$, $V^{\wedge}(k)$

113. $\underline{U}^{\wedge}$, $\underline{V}^{\wedge}$, $\underline{S}^{\wedge}_{\underline{F}}$, $F^{\wedge}(t,j,k)$

CHAPTER VI

Page:

116. $F_1^{\wedge}(t,k,j)$, $F_1^{\wedge}$, $w_2(\underline{y})$

117. $F_2^{\wedge}$, $F_2^{\wedge}(t,j,k)$

118. $G(t,m,\underline{x})$, $U^{\wedge}(t,m)$, $V^{\wedge}(t,m)$

119. $G(\underline{x})$

125. $w_{i,k}^{\wedge}(t)$

126. $w_{i,k}^{\wedge}(t;\underline{x})$

134. $\underset{\sim}{L}$, $\underset{\sim}{L}^{\wedge}$, $E(t)$, $E^{\wedge}(t)$

135. $f(\underline{\xi})$, $(\cdot,\cdot)_t$

136. $\underline{\xi}_k$, R_k, $g(\underline{\xi})$, $g_k(\underline{\xi})$, $\Phi_k(\underline{\xi})$, $\gamma(k)$

137. T_k, $M(w)$, E_k, $\Phi(\underline{y})$